URBAN ENERGY SYSTEMS

About the Author

Dr. B. Sudhakara Reddy completed his doctorate at the Department of Management Studies, Indian Institute of Science, Bangalore, India. Since 1991 he has been working as an Associate Professor at the Indira Gandhi Institute of Development Research, Mumbai. His books on "Planning for Demand Side Management in the Electricity Sector" and "Sustainable Regeneration of Degraded Lands" were published by the Tata-McGraw Hill Publishing Company Ltd., New Delhi. He published around 40 papers in reputed International Journals. He is the Guest-Editor for the *International Journal of Global Energy Issues, International Journal of Environment and Pollution and Energy Sources*. He is in the Editorial Board of the *International Journal of Sustainable Development*. He has served as consultant for the United Nations Development Programme. Dr. Reddy's work is concentrated mainly on Energy Efficiency, Energy Environment nexus, Sustainable Development and Renewable Sources of Energy. He serves on the Maharashtra State Planning Board as a member (Energy).

URBAN ENERGY SYSTEMS

B. Sudhakara Reddy
Associate Professor
Indira Gandhi Institute of Development Research
Gen. Vaidya Marg, Goregaon (East), Mumbai 400 065

CONCEPT PUBLISHING COMPANY, NEW DELHI-110059

ISBN : 81-7022-681-3

First Published 1998

Published and Printed by:
Ashok Kumar Mittal
Concept Publishing Company
A/15-16, Commercial Block, Mohan Garden
New Delhi-110059 (India)
Phones : 5648039, 5649024
Fax : 091-11-5648053

To the memory of

my late grandmother

SMT. RUKMINAMMA

who inspired me to be

what I am today

FOREWORD

Urbanisation is a global phenomena. Its pattern and characteristics depend on many factors such as regional differences.

Urbanisation invariably results in increased resource utilisation. Such resources, however, are generally limited. The situation forces decision makers to identify efficient patterns of resource use. Consequently, the resource constraint has far-reaching implications on national economies, particularly those of developing countries.

The proliferation of urban metropolises accompanied by their insatiable demand for limited resources has highlighted the urgency and importance of studies on urban systems. The oil crises of the 1970s and their impact on national economies underscored the need for understanding urban energy systems and evolving planned approaches to them. These studies have received greater impetus and appreciation because of the growing awareness of environmental issues and the challenge of sustainable development.

Analytical approaches to study urbanisation and the associated phenomena are necessary in order to improve planning and policy decisions. There is a need for empirical research of energy demands based on an end-use analysis of different energy carriers. The fuel-cycle approach towards end-use analysis—involving the tracking of the flow of energy from source to service— is a very effective way of studying the complete energy system. The approach includes sources, energy management systems and end-uses. Shifts in fuel/carrier usage are an important characteristic of urban systems in developing countries where there is stratification of income and a diversity of fuels/carriers for the same energy tasks, utilisation and income levels.

This book is an outcome of the research work carried out by B. Sudhakara Reddy during his tenure as my doctoral student at the Indian Institute of Science, Bangalore. It takes a close look at Bangalore, one of the fastest growing metropolises in India and presents a detailed analysis of its energy system as it evolved over a decade. It starts with detailed empirical data and statistical analysis of the usage of various energy fuels/carriers and shifts between them. The study shows income-dependent shifts in energy utilisation patterns from traditional to modern energy carriers such as petroleum products and electricity. It then goes on to the formulation and testing of analytical models. Policy implications are also discussed.

Bangalore is a typical example of a rapidly expanding metropolis in a developing country. Hence, the study has implications that extend beyond Bangalore. In particular, it throws light on different aspects of urbanisation and the associated energy systems in developing countries. The general approach developed in this book can be adopted to other urban concentrations, particularly in developing countries.

Finally, it should be mentioned that Sudhakara Reddy has demonstrated that through dedication and commitment, work of major significance can be achieved. Apart from its intrinsic interest, I hope that this book will be of use in the formulation and implementation of policy and programmes for urban systems.

Professor Amulya K.N. Reddy
President, International Energy Initiative
Bangalore.

PREFACE

I do not know what I may appear to the world,
but to myself I seem to have been only a boy
playing on the sea-shore, and diverting myself
in now and then finding a smoother pebble
or a prettier shell than ordinary,
whilst the great oceans of truth
lay all undiscovered before me.

—Sir Isaac Newton

As the world moves into the 21st century, it is expected that more than half of its population would live in urban centres. The problems of such a world with a massive urbanisation, particularly in the areas of energy and environment, will be daunting, and are feared to shift to the arena of the developing countries. Some claim that most of the energy resources of the world are utilised by the developed countries only. Even though it is largely true, urbanisation is a global phenomenon characterised by varied patterns and is dependent on many factors. Regional differences play a major role in determining its extent. Urbanisation often results in an increased resource utilisation. But, such resources are generally limited in nature forcing the decision makers to find efficient use patterns which in turn have immense implications for national economies, particularly those of the developing countries.

The importance of studies on urbanisation systems has been recognised due to the growth of urban centres accompanied by a rapid utilisation of limited resources. The oil crisis of the 70's and its impact on national economies brought into a sharp focus the need for understanding and planned approach towards urban energy systems. With the recent awareness about environment and sustainable development the need for these studies has received a greater impetus and appreciation.

Analytical approaches to study urbanisation and related phenomena are necessary in order to improve the planning and policy decisions. There is need for an empirical research of energy demands based on an end-use analysis of different energy carriers. The fuel cycle approach towards the end-use analysis is a very effective way of studying the complete energy system including various sources, energy management systems and end uses. Shifts in fuel usage are important indicators of the economic state of a particular urbanising system in the developing countries where exists diversity of fuel utilisation and income levels.

In this book we take a close look at the different aspects of urbanisation and related energy systems as applied to the developing countries. We consider specific cases and present both analytical models and statistical/numerical data to show the interaction among various energy systems in representative urban settings. As a typical example of a rapidly urbanising centre in a developing country we consider the case of Bangalore city and present a detailed analysis of its energy system for a period of 10 years. The study shows shifts in the energy utilisation patterns with income levels, from traditional to modern fuels like petroleum products and electricity. This type of study would be of an immense use for policy formulation and implementation. The general approach developed in this book can easily be adapted to other urban concentrations, particularly in the developing countries.

Many individuals and institutions have helped to make this volume possible. Apart from my own, the single most contribution came from Prof. Amulya K. N. Reddy, my Ph.D supervisor [President, International Energy Initiative (IEI), Bangalore]. He not only suggested this interesting problem but also the methodology to attack it. He argued logically point by point compelling me to clarify and integrate concepts on which this book was based. Despite severe constraints on his time, he meticulously went through several drafts reading line by line suggesting deletion and additions. It is in a large measure his book. I will ever be grateful to him.

I can never forget the help extended to me by Sri K. Sreenivasa Rao, Dr T. Srinivas and P. Balachandra of the Indian Institute of Science, Bangalore, and Dr Gladys D. Sumitra, and Ms Antonette D'Sa of IEI who showed keen interest and constantly helped me through all stages of the work. My sincere thanks to them. I am also indebted to Prof. Kirit S. Parikh, Director, and Prof. Jyoti K. Parikh, Head, Energy Division, Indira Gandhi Institute of Development Research, Mumbai, for their advice and comments and to Sri. J. C. Shah, former Chairman, Gujarat Electricity Board, for critically going through the manuscript. The secretarial help provided by Sri. Mahesh Mohan, IGIDR, is thankfully acknowledged.

A study of this nature which involved collection of massive data would not have been possible without the willing cooperation and assistance of a large number of people from different agencies and organisations. Many were kind enough to answer my questions during the course of the investigation. My sincere thanks to all of them.

My parents, wife Lalitha, sons Sandeep and Siddarth and nephew Murali smilingly accepted my prolonged absences from home. I am more than duly thankful to all of them for this forbearance.

Last but not the least my revered grandmother Smt. Rukminamma, had been the true guiding force behind my endeavours. Though living in a remote village she had the wisdom to foresee the great strides made in the field of education and encouraged me to pursue higher studies. I am ever greatful to her.

Mumbai **B. Sudhakara Reddy**
June, 1997

CONTENTS

1

INTRODUCTION

For a brief space it is granted to us,
if we will, to enlighten the darkness that surrounds our path...
We press forward, torch in hand, along the path.
Soon from behind comes the runner who will outpace us.
All our skill lies in giving into his hand the living torch,
bright and unflickering, as we ourselves disappear in the darkness.

– Havelock Ellis

Urbanisation is a global phenomenon. Both economically advanced and developing countries experience this phenomenon though characterised by different patterns of urbanisation. Such patterns involve a high level of availability and utilisation of resources, interaction among urbanised and non-urbanised centres and avenues for economic prosperity. However, economic growth and expansion can be considered as common factors.

Economic growth is a major policy objective especially for most of the developing countries. In pursuing this goal, these countries place an emphasis on industrial development which in turn has resulted in rapid urbanisation. Such an urbanisation brings in its wake a perennial movement of people from villages to towns and cities which gives an impetus to the growth of metropolitan areas. The increase in urban populations results in asymmetric patterns in resource utilisation, land use pattern, transportation, etc. Production and consumption of goods and services are accounted for mostly by urban centres because of the concentration of industrial and commercial establishments, and middle- and high-income households.

It is hardly an exaggeration to say that with increasing urbanisation, energy has moved to the very centre of public concern. As the population expands and people strive for better standards of living, the per capita energy consumption rises manifold. It is not surprising, therefore, that those who advocated "limits to growth" have been ridiculed for years as impractical and opponents of development. However, in recent years, there seems to be a realisation of the negative impact of growth, particularly in the area of "urban development".

The availability of *energy*, its utilisation and its impact on the *environment* and *economy* are important considerations for any urban region. The interconnection among these suggest that all these crises were symptoms of some common fault that lay deep within the design of the energy system that governs how energy is produced, transported, distributed and used. Thus, it is obvious that the idea of studying all these aspects of energy utilisation is more than ever in need of support, as the advances of science and technology multiply the hazards.

Urbanisation and Energy

Urbanisation involves many structural changes throughout the economy which have important implications for energy utilisation. Urbanisation increases per capita energy consumption and also leads *inter alia* to shifts in the overall pattern of fuel/energy carrier consumption. These carrier shifts are stimulated due to an increase in the monetary incomes of households and also by the availability of convenient and efficient energy carriers such as kerosene, LPG (liquefied petroleum gas), and electricity (Hoiser and Dowd, 1987; Reddy and Reddy, 1994). In addition, people from rural areas accustomed to firewood, cow-dung and agricultural wastes for cooking/heating switch over to modern energy carriers on migration to urban areas due to an increase in their income and the relatively easy availability of these carriers. Hence, the share of non-commercial fuels such as wood and charcoal, in the total energy consumption, has declined rapidly, from 67 per cent in 1953-54 to 35 per cent during 1991-92 [Anon, 1993].

The first period of concern regarding energy was in the 70s due to the Organisation of Petroleum Exporting Countries (OPEC)-induced rise in oil prices. This led governments to impose restrictions on energy consumption levels. However, none of these measures had much impact. The use of energy carriers like petroleum products and electricity continued to increase. There could be many reasons for this. Research studies conducted to assess the implications of energy prices have not critically examined the process of energy consumption (direct or indirect). Also the issue of the impact of energy consumption on forests, environment, transportation and the economy as a whole were hardly touched upon. This can be done only if a whole range of issues including the source of energy, the mode of transportation and the pattern of energy use are examined simultaneously. The second period of concern was in the 90s when the environmental dimension was added to energy consumption, for which a complete rethinking of our pattern of energy use including transportation, fuel switching, etc., is needed. In a sense, the threat of global warming can force the energy planners to re-examine the energy consumption patterns and plan for a *sustainable development.* A city with an efficient system of public transport, industrial processes and technologies has fewer environmental effects that are presently encountered than a city with an inefficient industry and services along with private automobile dominated transport.

The study of the urban demand for energy has to take into account all the above considerations. It should also be addressed as part of the national development plans. However, this cannot be done without an understanding of the inflow of various energy carriers and their pattern of utilisation in urban conglomerations. The coexistence and utilisation of several energy carriers in a stratified city is also an interesting study. In particular, it is important to understand the energy carriers used by the different sections of society and their end uses. Finally, it is necessary to examine whether the urban energy demand can be satisfied in a sustainable way and whether there are technological alternatives. These, then, are the various inter-related issues that motivate the study of an urban energy system.

One may wonder at the significance, relevance and utility of an exercise such as *Urban Energy Analysis*. Convincing answers can be found if one examines issues such as: what is the relative deprivation of energy at various levels of economic hierarchy? What are the carrier choices in the household sector and how they change according to various economic and social factors? What are the factors which are unique in an urban centre and are to be considered in the formulation of strategies and policies for the energy sector as a whole? An attempt is made here to examine these aspects and indicate the policy measures in the background.

Earlier Approaches

One of the most surprising aspects of the literature on an urban energy analysis is the scant attention paid to non-commercial energy sources like firewood and charcoal. There were a large number of studies on the demand for energy conducted in India and elsewhere (Halvorsen, R, 1975; Garbacz, 1983; Giuseppe and Ruggeri, 1983; Schipper, L, 1983; Garica, Manegdeg and Raneses, 1994). Most of them deal with commercial energy consumption, electricity consumption in particular. There has been little attempt to reveal the pattern of consumption of all energy carriers in a city (exceptions are: NCAER, 1958; Newcombe, 1975; McGranahan and Taylor, 1977; Schipper, 1983).

The urban energy analysis approach is a recent phenomenon. It started in 1974 in the form of a public discussion among scholars, which concentrated on Hongkong, Java and India. Studies on the energy analysis of the Third World cities were conducted by the NCAER (1958), Newcombe (1975), McGranahan and Taylor (1977), Meier *et al* (1978), Bowonder *et al* (1986) and Pongaspapich *et al.* (1994). Of these, the most thorough study on urban an energy analysis is that for Hongkong city by Newcombe. He studied the energy consumption in various sectors of Hongkong and found significant influences on pollution and changes in the microclimate as a result of an intense energy use.

The first urban energy consumption study in India was done in 1958 by the National Council for Applied Economic Research (NCAER) for Mumbai, Chennai, Delhi and Calcutta. This study sought to reveal the pattern of consumption of different energy carriers in different sections of society and the determinants that influence energy carrier consumption. The study also estimated the per capita consumption and concluded that income, family size, price and availability of the energy carrier and its substitute are the most important factors that determine the choice and consumption of an energy carrier.

Many other studies (Taylor, 1975; Meir *et al.,* 1978; Dunkerley, 1986) have considered the pattern of energy consumption in urban areas. These studies show that petroleum products and electricity were the main energy carriers even though the biomass fuels (firewood, charcoal and agricultural wastes) were being used to a greater extent. This energy was consumed mainly by industrial transport, commercial and residential sectors.

Energy consumption in different income groups was estimated by McGranahan and Taylor (1977); Hughes-Cromwick (1985); Nair and Krishnayya (1985); and Pongasapich and Wongsekiarttiart (1994). They disaggregated the energy demand by income group and showed that family income, family size, appliance ownership and price of the energy carrier are the major determinants of household energy consumption. With the exception of Taylor (1975); Meir (1978) and Dunkerely (1986) the focus of most other studies has been rather narrow, i.e. one can say, uni-dimensional. They confine their studies to the residential sector and even within this section only on electricity consumption.

A New Paradigm for Sustainable Development

The per capita energy consumption in an urban area is more than double that in the rural parts and the urban consumption is mainly through petroleum products and coal which are non-renewable. Further, as the trends suggest, the urban population will double by the end of this century and resource conserving energy designs are crucial. Efficient and flexible energy-use patterns are unlikely to evolve from the energy-intensive ones without policy intervention. This may be because of inertia, market imperfections and the fact that fuel shifts are not given proper attention by the urban planners. The environmental implications of energy consumption are not yet understood properly. Thus, the case for including energy carrier shifts, efficient utilisation of energy and energy-environment interactions in urban planning is both compelling and urgent.

Until recently, growth in energy use was equated with economic and social development. However, it has long been recognised that energy is not an end in itself; it is a means of providing services like illumination, space heating, conveyance, etc. As with the case of other resources like land, potable water and clean air, energy is exhaustible. It is also being distributed unevenly and will become increasingly difficult to extract and deliver it in the future. Finally, the transportation and utilisation of energy systems are causing an irreversible damage to human and natural environment. However, much of the information required to deal effectively with these issues is not available. An obvious reason is the absence of statistics on an end-use analysis of different energy carriers and efficiencies with which they are being consumed. So, even at this late date in the history of the energy scene, we have no comprehensive urban energy plan to describe the system, let alone correct the serious faults in the distribution networks and consumption patterns.

Urban energy has been entirely ignored in the planning process. Sufficient data is not available either on the pattern of consumption or on the changing energy needs of the urban people. It is necessary to study urban energy consumption in depth and model the fuel shifts. Many urban policy-makers and planners continue to ignore these issues or pay very little attention to them. It could not be because they are unaware of the implications of energy carrier shifts but somehow they seem to be biased in favour of supply-based approaches rather than demand-side management. An impression is gaining ground that energy efficiency and optimal allocation of resources to the various sections of society will conflict too much with social, economic and environmental goals of some vested interests. These are serious issues that need a thorough analysis and interpretation.

Since energy use is associated with resource depletion, the approach taken will have to be coordinated with other aspects of sustainable development, so that the broad goals and values of humanity can be realised. Interpretation of energy considerations in any urban planning process is bound to be constrained by the lack of information on various activities. Given the link between energy and environment, the need to use resources efficiently and the fact that income influences the energy carrier choice of people should not be ignored in urban energy planning. Energy use in a stratified society is different from that in an egalitarian one. State planning tries to reduce inequalities in energy use between various groups of people but cannot remove them. This is because resources are allotted on the basis of the strength of the competitive groups. Ideally, the process of planning allocates energy resources optimally among all the groups while in reality those with political and economic power influence the planning process. As commented by Saunders (1980), "Planning does not serve public interest because there is no such thing as public interest."

Even though there exist some models and approaches of urban planning for the developed countries, they cannot be applied to the developing countries. This is because of the variations in the urban areas of the developed and developing countries which differ in population growth, levels of income and end-use services. It is the paucity of relevant data and increasing need for models of urban areas in the developing countries that has necessitated the present work. Recently there has been a shift in the energy planning paradigm, from considering *energy as a producer to a service provider* which has brought these issues to the forefront. This book has been strongly influenced by that paradigm shift made popular by Prof. Amulya K. N. Reddy (Reddy A.K.N. *et al.*, 1990).

Instead of considering the role of the government to remedy the problems facing the urban people it is desirable to study the phenomenon, pattern of energy consumption in a stratified society and the beneficiaries from the existing energy policies. It is also apt to examine the chaos created by the insatiable appetite of the high-income urban consumer for energy and suggest appropriate policy measures to remedy the situation.

Although the proposed approach to urban energy analysis shares many features of the earlier studies, it does not look at energy consumption in isolation but views it from a larger urban perspective. It is the energy flow and end-use orientation along with the modelling of energy carrier substitution that serves as a point of departure from earlier studies. It identifies the issues in this connection and describes ways in which the energy analysis approach can be adopted for a sustainable development of the urban agglomerations.

With this broad perspective, Bangalore, a rapidly developing city in southern India, was selected to study its energy system. It was chosen for several reasons. Compared to other Indian cities, it has a cosmopolitan character, has the fastest growth rate (78 per cent between 1971 and 1981), depends only on automobile transport and, more importantly, consumes substantial quantities of non-commercial fuels like firewood and charcoal. It offers several pointers for further application of the *urban analysis approach* and *energy carrier substitution models* developed in this document. Applying this approach on a nation-wide scale is difficult, so this urban centre was chosen to serve as a reference case study and an indepth analysis was carried out. We are seeking to generalise, but are aware that the differences in the urban demographic and other characteristics are impediments. An indepth case study of this nature is bound to throw up some new factors or eliminate a few existing ones. Either way, the methodology and models developed here have to be replicated for other urban settings. This gives all the learning components, just as wisdom dawns after knowledge is acquired.

2

URBAN ENERGY SCENARIO

There is no substitute for energy; the whole edifice of modern life is built upon it
Although energy can be bought and sold like any other commodity,
it is not "just any other commodity" but the precondition of all commodities,
a basic factor equally with air, water and health.

– Schumacher

Urbanisation is a direct outcome of the ways in which different sections of society are organised. It has been argued (Gilbert, A and Gugler, J., 1980) that cities serve only the elite groups even though they consist of people with different characteristics and tastes. This means that egalitarian societies cannot produce cities. A combination of infrastructure development, industries and communications creates and sustains urban centres. Urbanisation picked up momentum during the Industrial Revolution. In the developing countries, it started only after the Second World War. Starting in 1950, the pace of urbanisation has increased dramatically. Between 1950 and 1990, the world's urban population has doubled, with the major cities expanding at a rate of 3 to 5 per cent per year.

The process of urbanisation is accompanied by major economic and structural changes throughout the economy. Urban areas reflect the general orientation of the region *vis-a-vis* concerned with international trade. Manufacturing, trade, commerce, financial services and transportation are the major employment avenues in urban areas. According to Castells (1977), urbanisation can be characterised by

- a link between industrial production and urban growth;
- a population unrelated to the productive level of the system;
- a strong imbalance in the urban network in favour of one predominant area;
- a lack of jobs and services for the new urban masses; and consequently
- a reinforcement of the ecological segregation of social classes and a polarisation of the system of stratification as far as consumption is concerned.

Urbanisation in developing Countries

About the Third World urbanisation, Murphy (1980) says,

> "... by the time of independence, national life had come to centre on western countries to an irreversible degree as each Asian country had responded to western stimuli and altered its own outlook. Its world had been refocussed on western manipulations but ultimately became the breeding ground for the apexes of new Asia. These apexes were to develop into primate cities which would dominate national settlement patterns. Into these cities, would eventually move large numbers of rural migrants, their lands alienated by intrusion of capitalist enterprise, their numbers swollen by lower mortality rates".

In terms of urbanisation levels, India is one of the least urbanised countries in the world. Between 1951 and 1991, the level of urbanisation has increased only by 8.4 percentage points. However, if the size of the population is considered, then, India has the second largest urban population in the world (217 million). Thus, the significance of urbanisation can be measured in terms of population rather than the level of urbanisation (Table 2.1).

Table 2.1: Urbanisation levels in India (1951-1991).

Description	Year					
	1951	**1961**	**1971**	**1981**	**1991**	**Change in per centage points**
Urban areas (no.)	2843	2365	2590	3378	3768	
Population (million)	62	79	109	160	217	
Urbanisation level	17	18	20	23	26	8
Percentage share						
Class I (UA)	3	5	6	7	8	6
Class II	3	6	7	8	10	6
Class III	12	19	23	23	26	14
Class IV	22	32	33	32	31	10
Class V	40	31	25	23	20	-20
Class VI	20	8	6	7	5	-15
Total	100	100	100	100	100	
Share of urban population						
Class I (UA)	45	51	57	60	65	21
Class II	10	11	11	12	11	1
Class III	16	17	16	14	13	-3
Class IV	14	13	11	10	8	-6
Class V	13	7	5	4	3	-10
Class VI	3	1	1	1	0	-3
Total	100	100	100	100	100	

A significant observation made from Table 2.1 is that more than two-thirds of the urban population lives in urban agglomerations (UA) (regions having a population greater than one million) whereas their share in the total is only 8.2 per cent. Between 1951 and 1991, UAs increased by only 5.5 percentage points while their population showed an increase of 20.6 percentage points. The percentage shares of Class IV to VI towns showed a declining trend (Urban areas are classified as follows: Population for Class I = > 1 million; Class II = 100 thousand to 1 million; Class III = 50 thousand to 100 thousand; Class IV = 20 thousand to 50 thousand; Class V = 10 thousand to 20 thousand; and Class VI = < 10,000).

In 1950-51, the contribution of urban India to the net domestic product was 29 per cent which rose to 41 per cent in 1990-91 (Anon, 1992). This figure is likely to reach 60 per cent by the turn of the century. Thus, on the one hand, urban areas are considered as "engines of growth", creating wealth and employment and, on the other, they consume significant quantities of resources at the expense of rural areas and thus degrade the environment.

Urbanisation and energy

The dynamic interaction between urbanisation and energy has been of considerable interest since the oil crisis of the 1970s (Anderson, K. P., 1973; Baxter, R.E. and Rees, R., 1978; Daxiong Qiu *et al.*, 1994). In an urban environment, land-use pattern, transportation, spatial structures, industrial location, etc., affect the levels of energy consumption.

The nature, availability and pattern of utilisation of energy sources influence urban society. Historically, energy transition - from dispersed organic sources through wind and water power to the exploitation of fossil fuelscan be linked to the major shift in human settlements. This shift resulted in the development of urban/industrial centres. Even if energy is not considered the prime cause for this major social change, it has undoubtedly played an important role.

If energy can be considered as a permitting factor in the process of urbanisation, there is little doubt that it has exerted a profound influence on the location and direction of urban development. Energy supply networks and increased personal mobility through efficient modes of transportation have removed locational constraints for industries and thus provided permitting factors for urban growth. In short, energy supply and its distribution networks are among the important factors that propell urban growth.

Just like in any urban agglomeration, a large portion of energy consumption in urban India is by electiricity and petroleum products like kerosese, LPG, motor spirit (petrol), high-speed diesel, etc. However, significant quantities of firewood, charcoal and coal are also being consumed.

A Case Study of An Urban Centre in India

Karnataka is one of the more urbanised States in India with 30 per cent of its population living in urban centres (Anon, 1991). However, the pattern of urbanisation is not uniform with variations in the intensity, rate of urbanisation and spread in various sub-regions. Bangalore, the capital of Karnataka State, accounted for 46 per cent of the total urban population. Another significant feature of urbanisation is the correlation between the size of the urban area and its growth rate; the bigger the area the larger its growth rate. For example, Bangalore achieved a growth rate of 7.6 per cent p.a. whereas other urban centres showed a growth rate of 4.9 per cent.

Karnataka has been a pioneer in power generation having started as early as 1902. Electricity was first supplied to the city in 1905. The installed capacity was 3,166 mW during 1989-90 and the power generation was 14,155 gWh. As in any urban agglomeration, a large portion of Karnataka's energy consumption is through electricity and petroleum products like kerosene, LPG, petrol, diesel and fuel oil. However, significant quantities of firewood, charcoal and coal are being consumed. Bangalore's peak demand was just 25 mW in 1951 which rose to 550 mW in 1985 and is estimated to go up to 1,200 mW by the turn of the century. Bangalore city accounted for around one-fifth of the total electrical installations, energy sold and energy entitlement of high tension (HT) industrial connections, and a quarter of the total connected load and contract demand of HT connections; it has a lion's share of 50 per cent in the All Electric Home (AEH) with a connected load of 15-ampere installations. This is not in line with the egalitarian distribution of an essential item. What is more obvious is the fact that this inequality has arisen for an item whose distribution is managed by a public utility and not dictated by market forces (Table 2.2).

The big concession given to electricity consumption has not checked the firewood consumption. A study of firewood consumption and supply has revealed that the daily consumption of firewood by households was 970 tonnes, accounting for 3.5 per cent of the State's firewood consumption. In the case of petroleum products, Bangalore accounts for more than 60 per cent of the light diesel oil supplies to Karnataka as a whole, about 45 per cent of motor spirit, LPG and kerosene, 37 per cent of furnace oil and 22 per cent of high-speed diesel. Transportation in Bangalore is dominated by road transport where automobiles and trucks consume about 90 per cent of the total transportation energy out of which 80 per cent accounts for passenger transport. Personal transport, particularly two-wheelers, accounts for 60 per cent of the total vehicle kilometres and their share in the energy use accounts for 30 per cent of the total. Bangalore city accounts for 40 per cent of motor vehicles in the State and 96 per cent in the district. The rapid growth in passenger vehicles, particularly two-wheelers (15.41 per cent p.a.), is partly due to the inadequacy of public transport and partly due to the

Table 2.2: Power development in Karnataka

Item	Unit	1989-90
Installed capacity	MW	2645
Hydel	MW	2245
Thermal	MW	420
Power generation	GWh	11110
Plant load factor	Per cent	77
T & D distribution network	Ckt.km	366058
T & D losses	Per cent	21
Power requirement	GWh	18275
Power availability	GWh	13967
Power surplus (+)/ deficit (-)	GWh	-4308
Power surplus (+)/ deficit (-)	Per cent	-24
Power consumption (utilities)	GWh	11978
Domestic	Per cent	15.3
Agriculture		32.2
Industry		49.0
Commercial		2.0
Others		1.5
Per capita consumption	KWh	252
Profit/loss of State Electricity Board	Rs/million	-840

substantial increase in the city's population. This has not only increased the total fuel consumption, but also the number of passenger kilometres. The result is an increasing density of vehicles and environmental pollution (Table 2.3).

Table 2.3: Energy carrier consumption in Bangalore and Karnataka (1989-90).

Energy carrier	Bangalore	Karnataka	Share (per cent)
Firewood (million tonnes)	0.49	14.00	3.54
Kerosene (million litres)	0.18	0.41	44.81
LPG (million tonnes)	0.03	0.07	46.51
Motor spirit (million tonnes)	0.19	0.41	46.06
HSD (million tonnes)	0.30	1.35	22.06
LDO (million tonnes)	0.02	0.03	63.64
Furnace oil (million tonnes)	0.08	0.21	37.37
Electricity (gWh)	2290.00	9910.00	20.37

The energy sector of any place is a product of its demographic variables such as area, population, number of households, commercial and industrial establishments, topography, power infrastructure, etc. The following sections present a panoramic view of the Bangalore Metropolitan region.

The Demographic profile of Bangalore

Bangalore is the capital of Karnataka State in south India. It is situated in the heart of the Deccan peninsula of India at 12^0 58' N Latitude and 77^0 33' E Longitude at an elevation of 3021 ft (900 metres). Recorded history dates its construction to 1537 by Kempe Gowda. During the early years, the city was a major textile and trading centre. The British, after their arrival in 1799, set up a new settlement outside the city called the civil and military station of Bangalore (Cantonment). To provide for its industrial growth, a railway station and a hydroelectric power plant, one of the earliest in the country, were built. The Bangalore City Corporation was formed in 1949 with the gradation of the erstwhile municipality. Since then the city has grown by leaps and bounds_from 69 sq km in 1949 to 500 sq km by 1990. Table 2.4 presents selected demographic indicators for Bangalore and Karnataka State.

Except for the relatively large share of nearly 20 per cent of urban Bangalore of non-agricultural workers, its share in population, households and workers is less than 10 per cent. The only indicator which exhibits a relatively high figure is the density of population. As per the 1991 census, it is 8,000 per sq km. Yet another glaring difference is the significantly higher workforce that Bangalore city is supporting in the non-agricultural sector; 95 per cent of this workforce is occupied in this sector as against 63 per cent in Bangalore district. However, on the whole, the population, households and workforce supported by urban Bangalore is comparatively lower when viewed in the context of its share in energy which will be shown later.

During 1984-85, Bangalore accounted for 18.5 per cent of the State domestic product and 29 per cent of the income generated from the non-agriculture sector. The share of income from registered manufacturing industries is 65 per cent. Bangalore accounted for 33 per cent of investment and 56 per cent of employment in large-, medium- and small-scale industries. Similarly the population of the city which was 0.8 million in 1951 grew to 3.6 million by 1991. While the population growth rate of Bangalore was just 1.8 per cent per year during 1901-1910, it was 5.8 per cent per annum during 1971-80, the highest growth rate in the country (Table 2.5).

Table 2.4: Selected demographic indicators of Bangalore (1990-91).

Item/Unit	Bangalore	Karnataka	Share
Area (sq km)	425.0	191891.0	0.2
Population (million)	2.9	37.0	7.8
Households (million)	0.57	6.4	8.9
Main workers (million)	0.95	6.23	7.0
Non-agriculture workers (million)	0.93	4.77	19.4
Total income (Rs.million)	44650.0	267784.0	16.7
Per capita income (Rs/month)	919.0	590.0	

Bangalore city has a road network of 1,500 km consisting of cement concrete and asphalt roads. The Corporation did not have any water supply and sewerage cell until 1964, when a special agency, Bangalore Water Supply and Sewerage Board (BWSSB), was set up. Bangalore water needs are met from three sources_the Cauvery, Thippagondanahalli and Hesaraghatta reservoirs. The city corporation covers an area of 500 sq km and has a population density of 8,000 persons/sq km. The major land use is residential (35 per cent), followed by roads (20 per cent). The remaining land is used for commercial, industrial and other purposes (Sreehari, 1987). The present and future patterns of land use indicate that between 1991 and 2000, the share of residential and commercial sectors will increase and that of transport, industry and parks will decrease (Table 2.6). The figures indicate that the land area provided for various services like roads is inadequate and may be insufficient to meet the growing needs of the city. Also increasing vehicle population and encroachment of footpaths by street vendors causes congestion of roads deteriorating the quality of traffic service. These traffic hurdles result in low travel speeds, frequent stops with acceleration and deceleration of vehicles, leading to increased fuel consumption and exhaust pollution.

Bangalore is also a major centre for institutions of higher learning and research, viz. the Indian Institute of Science (IISc), Indian Institute of Management (IIM), Indian Institute of Astrophysics (IIA), Raman Research Institute (RRI), National Institute of Mental Health and Neurosciences (NIMHANS), Indian Institute of Horticultural Research (IIHR), Indian Space Research Organization (ISRO), National Aerospace Laboratory (NAL), Bangalore University and a host of engineering and medical colleges.

Giant public sector undertakings like Bharat Electronics (BE), Hindustan Machine Tools Ltd (HMT), Indian Telephone Industries Ltd (ITI), Hindustan Aeronautics Ltd

Table 2.5: Population of Bangalore.

Year	Population	Growth rate (per cent)
1901	159046	
1911	189485	19
1921	237496	25
1931	306470	29
1941	406760	33
1951	778977	92
1961	1199931	54
1971	1653779	38
1981	2913537	76
1991	3628165	27

(HAL), Bharat Earth Movers Ltd (BEML), etc., are also located in Bangalore. Recently, with the economic liberalization of India, Bangalore has become the home for many well-known multinational corporations like Industrial Business Machines (IBM), Hewlette Packard (HP) and so on.

The point one should keep in mind is that the present study considers the consumption of direct energy by Bangalore city. However, the city also uses energy indirectly by utilising a large number of goods and services produced elsewhere. This indirect energy content may be quite high and it is worth assessing the same. However, for the present study, the discussion is restricted to direct energy only.

Table 2.6: Land use pattern in Bangalore (1989-90).

Type of use	1991	2001
Residential	34.36	37.54
Transportation	20.83	20.53
Parks, playgrounds, etc	10.09	9.90
Industrial	10.14	9.92
Agriculture	4.17	3.09
Commercial	3.49	4.18
Others (schools, public buildings, etc)	16.92	14.84
Total	100.00	100.00

With this background, the energy consumption pattern was examined for the city of Bangalore. The process of examination is referred to as the *fuel cycle approach*.

Fuel Cycle Approach

This approach describes the flow of energy from source to end-use (Fig 2.1) [Reddy, 1986]. At the supply end, the primary energy is in the form found in nature - trees for wood, a water-fall, an oil well, etc. This *primary energy* is converted into convenient energy carriers designated as *secondary energy*, for example, firewood logs, charcoal, kerosene, LPG, electricity, etc. Secondary energy, after transportation/transmission and distribution is delivered as *final energy* to consumers in the domestic, industrial, agricultural, commercial, transport and other sectors. The final energy delivered to consumers is converted through various end-use devices (stoves, lamps, furnaces, engines, etc.) into the *useful energy* that provides energy services (cooking, illumination, process-heating, shaft power, etc.) and satisfies the basic needs. There are losses associated with the conversion of one form of energy into another. These losses are particularly significant in the conversion of final energy into useful energy. All the

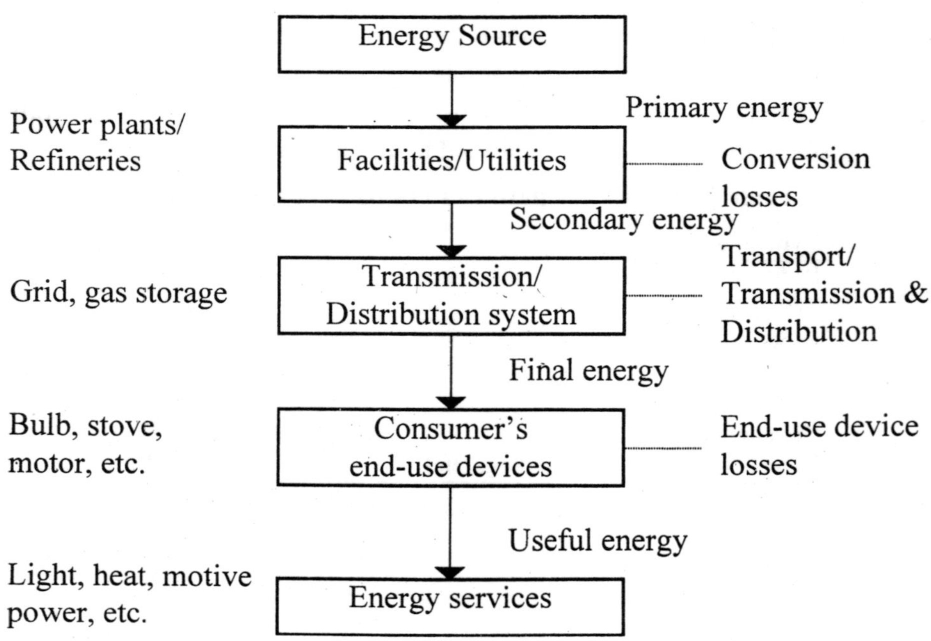

Figure 2.1: Fuel cycle approach.

energy carriers, viz. firewood, charcoal, kerosene, liquefied petroleum gas (LPG), electricity, petrol, diesel, fuel oil and coal are considered in the present analysis. The supply, transportation, distribution, consumption and end-use analysis of these energy carriers have been studied.

Efficiencies of Energy Flow

In many energy analyses, the efficiencies usually employed pertain only to the end-use of the devices. This method does not consider the amount of additional energy involved (such as diesel) for transportation/transmission of energy resources to the end-use point. Therefore, to analyse the flow of energy from source to end-use service, it is necessary to study the following efficiencies:

- End-use device efficiency (EDE): This is the efficiency of the device used for satisfying the demand of a particular end-use. This is used for calculating the useful energy.
- Simple energy efficiency (SEE): This is the overall efficiency of using an energy resource for meeting the demands of a particular end-use. It considers the loss of the energy resource at each stage, i.e. from the supply point to the end-use point. More specifically, this efficiency takes into account the loss of the energy resource during the processing from primary energy, transportation to the end-use point and losses in the end-use device. A significant characteristic of SEE is that it considers the loss of only the energy resource for computing it.
- Complex Energy Efficiency (CEE): This is an improvement over SEE. It is the overall efficiency considering the loss of the energy resource from the supply point to the end-use point. CEE also takes into account the amount of other energy resources spent while transporting the energy resource from one place to another. Specifically diesel spent on transportation will be taken into account in calculating the CEE. Obviously, CEE provides a clearer picture compared to SEE as it considers other significant energy losses incurred before deriving the end-use service using a particular energy resource.

In this book, however, only the end-use efficiencies have been taken into consideration due to lack of information on other efficiencies. Data were collected pertaining to the sources and supply of energy carriers, modes of transportation, channels of distribution and patterns of consumption in all sectors with an emphasis on the household sector. To collect this data, the agencies concerned were approached, their staff interviewed and sample surveys conducted where the population of consumers was large. For an analysis of the data, regression analysis was used wherever applicable.

Sources of data

Data on the consumption of various fuels/energy carriers were obtained from the organizations and agencies concerned. The number of trucks used to transport firewood and the quantity of firewood thus supplied were obtained from the five weighbridges in Bangalore. The number of railway wagons that supplied firewood and the quantity thus supplied were obtained from the zonal railway organizations. The quantity of firewood supplied to the consumers was obtained from the Forest Department. Interviews with the commission agents and truck drivers yielded information on where the firewood came from. The urban households do not have access to forests and hence they purchase firewood from the depots. The firewood is weighed at the time of purchase, the frequency of purchase normally being once a week.

The quantity of charcoal supplied to Bangalore was obtained from the Forest Department and the Firewood and Charcoal Dealers' Association. The quantity of LPG, and other petroleum products consumed in Bangalore was obtained from the oil companies, viz. Indian Oil Corporation Ltd, Hindustan Petroleum Corporation Ltd, Bharat Petroleum Corporation Ltd; and Indo-Burmah Petroleum Corporation Ltd. The details of electricity consumption were obtained from the Karnataka State Electricity Board. The supply of coal was obtained from the railways and the consumption details were obtained from Coal India Ltd, and the Karnataka Small Scale Industries Development Corporation.

Since most of the sources from which data were obtained are government or quasi-government offices/agencies, their records are subject to a periodic review or audit and hence they have been considered (to a first approximation) to be as reliable as other government data in India, such as the Census or National Sample Survey. As such, the data are recorded information. However, they are not the published ones and hence not cited. In the case of households, it is direct first-hand information.

Since each class of data is obtained from a unique source, inconsistencies have not arisen. Within themselves, they are consistent as displayed by further interpretation and analysis.

Household Survey

The consumption data obtained for various fuels/energy carriers indicated that residential energy consumption was the highest, accounting for about one-third of the total energy consumption. Therefore, it was decided to carry out an indepth analysis of household consumption. However, the consumption data obtained from the various sources reveal only the quantity of energy carriers consumed. They do not throw any light on the pattern of consumption by different sections of the society, the per-capita consumption for different end-uses such as cooking, water-heating and lighting and the determinants that affect the consumption, etc. Therefore, to get an accurate picture of the residential consumption, a sample survey was conducted.

Size of the sample

A sample of 100 is expected to give an estimate of the average consumption of an energy carrier per household with a standard error of <10 per cent (Maddala, 1986). However, the sampling theory indicates that, in large samples, the error variance can be controlled and stability in estimates can be obtained. Also, since the survey has to throw light on the consumption pattern of different types of energy carriers, the coefficient of variation changes widely. The following formula (Srivastava, *et al.*, 1980) was, therefore, used to determine the sample size:

$$n_p = (N_p) / N * n \text{ and } \sigma_m = \sum_{p=1}^{n} W_p^2 \left(\sigma_p^2 / n \right) \qquad (2.1)$$

where

n_p = required sample size

N = total number of households in the population,

N_p = total number of households in the pth cluster,

W_p = weight of the pth cluster = Np/N,

σ_p = standard deviation of the pth cluster, and

n = total number of clusters in the population,

σ_m = standard deviation of the mean

On this basis, for a city of about 600,000 families, it turns out that a sample size of about 1,000 is adequate to achieve a standard error of <10 per cent.

Method of sampling

A preliminary survey of the city households indicated that social and economic characteristics of human population appear to be related to household income. Some measure of these characteristics could be a useful variable in the design of the sample. As the division of the city on household incomes was not possible, geographical stratification was adopted under the assumption that households with similar incomes and similar habits of consumption tend to aggregate in the same locality. This is called cluster sampling. Clustering of households is necessary in order to reduce the cost and time of data collection to manageable proportions. Using this method, the survey was carried out in the following manner.

For operational reasons, the metropolitan city of Bangalore was divided by the Corporation into 120 geographical areas out of which 40 clusters were formed in a fairly convenient manner. Groups of areas were placed in the same geographical class with the objective of achieving some degree of equality in sizes (in terms of 1981 households) and obtaining identifiable boundaries. Thus each cluster consisted of a relatively homogeneous group of areas, in terms of income, family size, food consumption, etc. So it is reasonable to extend the homogeneity to energy consumption. Of the forty clusters, four were chosen based on their representative character to the total population, proximity and ease of data collection.

The sample was designed to yield a total of 60,000 households from the four clusters. In order to obtain a generally representative sample, 1,000 households were chosen (250 in each cluster) to obtain the characteristics of the households. The choice of households was random as follows. The household numbers were obtained from the City Corporation and are renumbered from 1 to *N*. The desired number of households, namely, 1000, were chosen by selecting random numbers until the required sample size was reached. The designed sample thus obtained was searched roughly for any bias, inconsistency and variance in terms of household income, with encouraging results. Thus each family in the city had the same probability of inclusion once the cluster was chosen.

It can be mentioned here that clustering of households is necessary in order to reduce cost and time of data collection to manageable proportions. But their representation of Bangalore is proved in the case of energy carriers, other than firewood and charcoal, where the data are available. The contiguous location of clusters was, to some extent, incidental but, in terms of similar incomes and habits, it is reasonable that they represent Bangalore well even in other terms as proved in the preliminary study.

Collection of data

The household survey was based on personal interviews. The questionnaire was a detailed one consisting of the name of the head of the household, family size, family income, type and quantity of the energy carrier used, etc. During the survey, the households were asked to enumerate the energy carriers used for different end-uses, viz. cooking, water-heating, lighting, etc. The pattern of end-use of each carrier was studied by disaggregating the consumption in the household according to cooking, water-heating, lighting, etc. Since the survey contains questions only on the total consumption of each energy carrier by the household (and not for a particular end-use), the disaggregation was done while analysing the results. Except in the case of electricity, other carriers are used only for cooking and water-heating and these habits hardly change with season (confirmed during the survey). Some members of the affluent families take a bath twice a day in summer but use electricity. The consumption of electricity is higher in summer because of the extensive use of fans, water-heaters, refrigerators and also, to some extent, bulbs and tubes.

Before the interview, the households were briefed about the aim of the survey which was to determine the pattern of consumption of energy carriers in the city of Bangalore. With regard to income, questions were asked in different ways to elicit more reliable information. When a household was unwilling to divulge its monthly income, the households were asked to indicate the income bracket they belonged to -- < Rs 1000, Rs 1000-2000, > Rs 2000, etc. The consumption of energy carriers was determined as usage per day, week and month. In the case of LPG, the number of days a cylinder lasted was ascertained. Electricity consumption data was obtained from the

Karnataka State Electricity Board after ascertaining the electricity meter number from the households. Seasonal variations are negligible for other fuels which was confirmed by the response of the households.

The income of the household was divided by the size of the family to obtain the per capita monthly income. The households were then categorised into seven per capita income groups: (i) < Rs.100, (ii) Rs.100–149, (iii) Rs.150–199, (iv) Rs. 200–249, (v) Rs.250–299, (vi) Rs. 300–349, and (vii) > Rs.350. Thus, the energy consumption patterns were studied as a function of the above income groups.

In later sections a close look has been taken at the supply, transportation, distribution, consumption pattern and end-uses of different types of energy carriers. The household survey data was used to gain an insight into the residential energy consumption.

3

ANALYTICAL FRAMEWORK

"To be sure, rational thinking and careful measurement belong to the scientific work, just as the hammer and chisel belong to the sculptor. But in both the cases they are merely the tools and not, the content of the work."

– Werner Heisenberg

Urban systems exhibit a variety of characteristics. This diversity arises due to variations in income levels, population growth, technology utilisation, energy consumption, etc. They influence the urban system and play an effective role in the modelling process. Urban models provide paradigms for understanding the structure of cities and are useful for policy purposes.

At present, there is a lack of systematic work on urban energy systems. This is a serious impediment in understanding them and ultimately affects their design. Construction of such a design/model can be used for prediction or projection, impact analysis and controlling and directing change in the urban energy consumption. The models that are developed should be small, comprehensible and easy to operate. The presence of a large number of energy carriers, influencing factors and their inter-relationships, and the long-term consequences of the decisions taken by the planners make the process of modelling very complex.

Cities in the developing countries exhibit phenomena that are different from those observed in Western cities. Their distinguishing features include the importance assigned to the role of traditional fuels and stratification of energy carrier utilisation based on income levels. Thus a typical model for an urban energy system in a developing country should consider such energy carriers as firewood, charcoal, coal, electricity and petroleum products. These carriers are consumed mainly in residential, commercial, industrial and transport sectors. Among these, the share of the residential sector in the total energy consumption is high (Dunkerley, 1980; Alam, M *et al.*, 1983). Unlike other sectors, there are many factors such as family income, size, etc., which play an important role in the energy carrier choice in this sector (Hirst, E, 1982; Garica, R. A. *et al.*, 1994). Thus, an analytical model for the residential sector is useful for understanding the overall picture of the urban energy system and has an immense application to policy formulation and intervention.

Many studies (Hoiser and Dowd, 1987; Reddy and Reddy, 1984) suggest that with a variation in the household characteristics, there will be a shift in the use of energy carriers. To examine how economic and other household characteristics influence energy consumption, the regression method is useful. Here we develop models to understand the mechanism that governs the energy carrier substitution process and study the factors that influence them.

Regression Analysis

The multiple regression method can be used to study the impact of different variables on energy consumption. It also allows an estimate the goodness of the fit of the equations and examines their merits. In doing so, the effect of the inclusion of a particular variable in the general fit of the equation and in the size and significance of the parameter estimates could be studied (Maddala, 1986).

Regression model

Let us consider an urban energy system consisting of carriers j (1 to N) whose consumption Y is effected by the variables X. The regression equation specifies a linear relationship between the monthly consumption of the jth energy carrier by the ith household, denoted by Y_{ij}, and the l explanatory variables X_{ij}. The regression model is specified by the following equation:

$$Y_{ij} = \sum_{l}\left(X_{ijl} * \beta_{jl}\right) + \varepsilon_i \tag{3.1}$$

where

Y_{ij} = monthly consumption of *j*th carrier of *i*th household

X_{ijl} = explanatory variables of the *i*th household using *j*th carrier for *l*th end-use.

β_{jl} = regression coefficient of *j*th carrier for *l*th variable

ε_i = error term

The explanatory variables considered as relevant in the above equation are (i) family income, (ii) family size, (iii) price of energy carrier, (iv) availability of carrier, (v) auxiliary fuel consumption and (vi) electrical appliance stock (in the case of electricity). Small fluctuations in the consumption of a particular carrier can be considered insignificant thereby making the model compact.

Model specification

The explanatory power of an equation is determined by the independent variables included in the model and the form of the relationship between them. Though the fit of the equation is important, it is not the only criterion to be used. It may be possible for an equation to have a high value of R^2 and insignificant coefficients of the explanatory variables. This may be due to the lack of explanatory power of the variables or there may be correlations among the variables. For this, two bases of comparisons can be adopted: the general significance level of the variables in the equation and the degree of multicollinearity present in the model. Therefore, (a) adequacy of the fit of the model and, (b) severity of multicollinearity were made by inspecting R^2, F test and the correlation matrix. Further, the "t" test was performed to assess the stability of the estimates.

Interpretation of parameter estimates

The usual interpretation of the parameter estimates is that they are partial rates of change. For example, by increasing the family income by one unit, the energy carrier consumption increases by a certain amount, all other factors remaining constant. Fundamental to this interpretation is the assumption that a variable can in fact be changed by one unit without affecting the other variables and that the model fit to the available data will apply when one of the variables thus changes. In this analysis, the data are the observations (not under the control of the analyst). So, whether or not energy carrier consumption would be increased by increasing the family income cannot be directly assessed. Rather, a more conservative interpretation is that the families with higher incomes are observed to have a higher firewood consumption. The sign of the parameter estimate indicates the direction of the relationship between the dependent and independent variables.

Total Energy Carrier Dependence Index

The usage of an energy carrier depends upon the income of a family. This income dependence of energy carrier usage can be explored through the total energy carrier dependence index (TEDI) which may be defined as follows:

$$TEDI = \frac{X_{ij} / \sum_{j} X_{lj}}{X_{l} / \sum_{l} X_{l}} \tag{3.2}$$

In other words, TEDI indicates the fraction of households in the lth income group using the jth energy carrier in the ith household.

A total energy carrier dependence index of unity means that the fraction of household totally dependent on a particular energy carrier in a particular income group is exactly the same as the category's fraction in the sample. Using TEDI, the dependence of a particular energy carrier in different income groups can be obtained.

Energy Ladder Concept

In the residential sector, households try to shift from one fuel to another if there is a change in their economic status (Reddy, 1994). With an increase in income, solid fuels (charcoal and firewood) give way to a liquid fuel (kerosene) which in turn can be displaced by gas (LPG) and electricity which are the most desirable energy carriers for households.

The energy ladder concept indicates that the pattern of energy use in different households varies with their economic status. Each step of the ladder corresponds to a different and more sophisticated energy carrier and the step to which the household climbs depends mainly on its income (Reddy and Reddy, 1983; Hoiser and Dowd 1987). While the low-income group households use firewood, cow-dung, agricultural wastes, etc., the middle-income groups use kerosene whereas the high-income households utilise LPG and electricity. The height of the step is determined by economic factors like the capital cost of the fuel-utilising device, price of the energy carrier, availability of the preferred carriers and household energy consumption.

Once a household decides to shift from one carrier to another, it has also to decide on the quantity of usage. This is because, with higher incomes, food habits change, more specifically the variety of dishes prepared increases, thus increasing the carrier consumption. On the other hand, if the economic status of the household decreases, either through a decrease in its income or through an increase in carrier prices, the household climbs down to a less-sophisticated carrier and consumes less quantity of fuel (Hosier and Dowd, 1987). To interpret the pattern of consumption of various energy carriers used by the households, two utilisation indices are defined.

Carrier utilisation indices

The first utilisation index, (UI_1) is for an income group and reveals the dependence of each income group on different energy carriers for an end-use. It also defines the fraction of households in a particular income group which use a particular energy carrier for a particular end-use.

Denoting X_{ijk} as the number of households in the *i*th household using the *j*th energy carrier for *k*th end-use,

$$UI_1 = \frac{X_{ijk}}{\sum_k X_{ijk}} = \frac{X_{ijk}}{X_{ij}} \qquad (3.3)$$

The second utilisation index (UI_2) is for an energy carrier and reveals the dependence of each income group on different energy carriers for an end-use. It also defines the fraction of households using a particular energy carrier which belong to a particular income group.

$$UI_2 = \frac{X_{ijk}}{\sum_j X_{ijk}} = \frac{X_{ijk}}{X_{ik}} \qquad (3.4)$$

The carrier utilisation indices indicate the variation in the contribution of different energy carriers to the energy mix of different income groups.

Energy Carrier Substitution

The fundamental assumption of the technological substitution is that there is an upper limit to the growth of a technology and the growth pattern follows a logistic path. According to this, each technology undergoes four different phases, viz. learning, growth, saturation and decline.

For each technology, there is always a learning curve during the initial stages. The growth phase is usually a logistic substitution, but it usually ends up before being fully saturated. Then the growth rate slows down and declines logistically (Marcetti, 1972). Therefore, it can be assumed that

- the new technology enters the market and grows logistically and also declines logistically.
- at any given time, one and only one technology will be in the saturation phase.

Thus, when the technology which is in saturation starts declining, the later technology gains acceptance and will in turn get saturated when a far more superior technology enters the market and this process repeats itself continuously. However, apart from the purely economic factors, other social factors such as family size, occupation of the head of the household, family traditions and availability of the carrier also play a role in the choice of the fuel (Alam, *et al.,* 1985). An attempt has been made to formulate mathematical models that provide a greater insight into the energy carrier substitution process.

Development of the carrier shift model

For the present study, firewood, kerosene, LPG and electricity can be considered as different energy carriers competing through their appropriate devices (stoves) for adoption by households. Firewood is replaced by an intermediate energy carrier, kerosene, which in turn is replaced by LPG or electricity. This process of substitution takes place as the income of the household increases. However, family size, occupation of the head of the family, tradition, availability and price of the fuel also play an important role in the selection of the fuel. In the case of occupation, if the head of the family is educated and employed, he tries to choose an energy carrier which is clean and can be used efficiently. Tradition and availability also play some role in the selection of a fuel. Therefore, other than household income, family size, occupation (employed or unemployed), tradition (preferring the fuel for traditional purposes), availability (available or unavailable) and price of the energy carrier are included as explanatory variables.

It can be argued that the likelihood of choosing a particular carrier depends on the economic and social status of the household and the decision here is discrete, i.e. selecting or not selecting. The problem, therefore, is of the discrete choice of the household for inter-fuel substitution where the household's characteristics play a major role. A multilogit model has been used here to explain the probability of selection of the fuel for cooking/water-heating by the household.

Formulation of the model

Regression analysis is a method of studying the relationship between dependent and independent variables. In such problems, all the variables involved take the values in some subset of real numbers. But, this method cannot be applied to problems where the dependent variable takes binary values, such as yes-no, present-absent, etc. Therefore, a technique called *logistic regression* has been applied to model the problem where the assumed underlying distribution of the error term is logistic.

This method is employed here to study the preference of fuels by households where the dependent variable *Y* denotes the preference or otherwise of a particular fuel with respect to another fuel (for example, wood against LPG) and is thus a logical variable. The values, one and zero, can be assigned to Y, depending on whether the fuel is preferred or not. The independent variable is the vector X denoting the characteristics of the household, i.e. size and income of the family, and other variables such as the price of the fuel, etc.

In the present analysis, the basic assumption is that the log-odds ratio (defined as the ratio of the probability of preference of the fuel to non-preference) is linearly regressive over the independent variables. The regression coefficients are determined using the Maximum Likelihood Estimation (MLE) over the sample households consisting of a set of n points in the form (X, Y).

Let $p(X)$ denote the probability that Y equals one when independent variables assume the value of X. Then,

$$p(X) = p\{Y = 1\} \qquad (3.5)$$

and

$$[1 - p(X)] = p\{Y = 0\} \qquad (3.6)$$

Thus, the odds ratio can be defined as:

$$p(X)/[1 - p(X)] \qquad (3.7)$$

and the log-odds ratio as:

$$R = \ln\,[p(X)/\{1 - p(X)\}] \qquad (3.8)$$

The logistic model assumes that the log-odds can be expressed as a linear combination of values of independent variables

$$X = \{X_1, X_2, \ldots\ldots\ldots X_k\}.$$

Thus:

$$R = \sum_{1}^{k} X_i \beta_i = X\beta \qquad (3.9)$$

where k is the number of independent variables and ß, the regression coefficients. The problem of interest is the estimation of parameters which denote the contribution of the ith household character to the log-odds ratio.

Before proceeding to the estimation of the model, in terms of application, the model can be interpreted as follows. Suppose, the problem of interest is the determination of the probability of any household preferring the energy carrier under study. Let $A = (a_1, a_2, \ldots a_p)$ be the vector of variables corresponding to a household (here, it is implicit that A need not belong to the sample studied, but its value must belong to the range of variable X). Substituting $X = A$ in equation (3.7),

$$p(A) = \exp.(A)/[1+\exp.(A)] \tag{3.10}$$

Thus, the analysis can be used to study the fuel preferential characteristics of the households.

Estimation of ß: Let the number of households in the sample be N. Corresponding to each household, we have the vector X_i ($i = 1,2,\ldots.N$) and the binary responses Y_i (which have independent Bernoulli distributions). In the present analysis, Ys denote the presence or absence of a particular fuel and the Xs, the household characteristics. Then, the distribution f, is:

$$f_i(Y_i) = p_i^{Y_i}[1 - p_i]^{(1-Y_i)} \tag{3.11}$$

where $p_i = p(X_i)$

Assuming that the selection of fuel by the household is independent, the joint distribution is the product of N such terms.

$$f = \prod_{1}^{N} f_i = \prod_{i=1}^{N} [p_i]^{Y_i} [1(1 - p_i]^{1-Y_i} \tag{3.12}$$

The log likelihood is, therefore, the sum of individual likelihoods.

$$\log f = L = \sum_{i=1}^{N}[Y_i \ln p_i + (1 - Y_i)\ln(1 - p_i)] \tag{3.13}$$

Now, since the probabilities are modelled in terms of log-odds ratio, equation (2.10) can be used to replace p_i.

$$L = \sum_{i=1}^{N}[Y_i \ln\frac{\exp(X\beta)}{1+\exp(X\beta)} + (1 - Y_i)\ln(1 - \frac{\exp(X\beta)}{1+\exp(X\beta)}]$$

$$= \sum_{i=1}^{N} [Y_i \ln(X_i \beta_i) - X_i \frac{\exp(-X\beta)}{1+\exp(-X\beta)}$$

$$= \sum_{i=1}^{N} Y_i \sum_j X_{ij}\beta_j - \sum_i \ln[1+\exp(\sum_j X_{ij}\beta_j] \quad (3.14)$$

Here L encompasses the effect of various parameters and may be used for optimisation purposes. The necessary condition for a function to be a maximum is that its first order partial derivatives must be zero. Therefore, differentiating with respect to ß, the equation becomes:

$$\frac{\partial L}{\partial \beta_j} = \sum_{i=1}^{N} \left[Y_i X_{ij} - X_{ij} \frac{\exp\left(\sum_{j=1}^{n} X_{ij}\beta_j\right)}{1+\exp\left(\sum_{j=1}^{n} X_{ij}\beta_j\right)} \right] \quad (3.15)$$

This is a system of n non-linear equations in j variables. These equations were solved using statistical subroutine package CO5NBF.

The parameter estimates indicate the effects of change in independent variables on the log-odds ratios, but not on the probabilities themselves. A non-zero estimate means that the percentage change in p_i and p_I are not the same. For any p_i, however, a zero estimate does not mean that the probabilities do not change. In particular, the effect on a particular p_i depends on the magnitude and sign effects on all other log-odds ratio.

Energy Carrier Competition

The results of the above model indicate that income is the most important factor which influences the energy carrier shifts. With increasing incomes, households switch over from one carrier and its associated device to another carrier and its associated device. Thus, the substitution of one carrier by another can be looked upon as a competition between two carrier-using devices for adoption by households so that the ratio of the number of households using the two devices is a function of the household income. With this perspective, a mathematical model for carrier substitution can now be developed to provide greater insight into the substitution process.

The proposed model describes the process of competition between different energy carriers competing through their appropriate devices (stoves) for adoption by households. Such a multi-level process can be called the process of replacement of an evolutionary type.

Technological substitution

A thorough understanding of the basic mechanism involved in the diffusion of new technologies is a prerequisite to any reliable assessment of the possible impact of technological developments on society in general. Also, long-term energy planning demands that we should have some idea of how a technological competition is likely to proceed in future. Through this one can foresee its impact on the demand for different energy carriers (Goldemberg *et al.*, 1988). In forecasting the technological competition and thereby the most likely events of the future, the curve fitting technique may not take into account several important factors that affect the economic and other decision-making processes. To overcome this, a competition model is developed to evaluate the factors quantitatively. The model is described and its application is illustrated in the case of competition of various energy carrier pairs.

In the present study, a model of energy carrier competition in the residential sector has been attempted not only with a view to understanding the mechanism that governs the substitution process but also to study the factors that influence it. Even though it is generally accepted that logistic equations follow the Lotka-Volterra equations (Lotka, 1956), it is an accepted fact that energy carrier competition is intimately related to economic considerations of households (Sathaye and Meyers, 1985; Cromwick, 1985; Daxiong *et al.*, 1994). With this in view, a competition model was developed to get a greater insight into the substitution process. This approach should provide a better understanding of the dynamics of the process of technological competition in the residential sector.

Most technological competitions can be described as a substitution of one technology for another. Each such substitution follows an S-shaped or logistic curve. Generally the substitution starts slowly, to overcome the initial problems and resistance, then it proceeds more rapidly as the competition between the old and new technologies grows stronger and the new technology gains advantage and then approaches saturation. Finally it declines when another technology gains acceptance and this process goes on.

In 1961, Edwin Mansfield (1961) developed a market substitution model. Blackman (1971) and Fisher and Pry (1971) have also developed substitution models. In these models, the independent variable is time rather than income. Instead of attempting to review the advantages and disadvantages of these models, we developed a model appropriate to the present competing energy technologies in households. The model is presented below and its application illustrated in the case of various pairs of energy carriers where a new technology competes with and gradually replaces the old one. However, in most cases the new technology does not completely supplant the old one which still continues to retain a portion of the household market.

Development of the model

The Lotka-Volterra equations describing the competing biological systems like predator - prey systems are used in the present study to model the competing energy technologies (for example, firewood, kerosene, electricity, etc.) through their respective devices. The arguments presented in Table 3.1 show that the comparison between the two systems is valid.

In the present case, we wish to study the competition of technologies in terms of the number of households associated with each of the energy carriers $X_1$and X_2. Here, the variation of X_1 and X_2 is with income I rather than with time t. In the absence of competition, i.e. when the systems are isolated, we expect a gradual saturation. Thus, we arrive at a non-linear variation of the rate of changes with the actual population (Pearl, 1935). The interaction with other technologies is assumed to sustain with increasing income. So, the technologies must oppose each other, but aid themselves (except for inherent decay factor $-d$). Thus, all the concepts of technological competing systems have analogies in the biological systems, to a major extent.

Consider two technologies, T_1 and T_2, competing in terms of the number of households X_1 and X_2, using them the medium of interaction being the household income I.

First, consider the possible variation of T_1 and T_2 in isolation., i.e. only one of them, say, T_1 alone existed. The simplest model is the one where the rate of change of X (either X_1 or X_2) depends solely on the current state of technology. In mathematical terms,

$$\frac{dX}{dI} = \Lambda X \tag{3.16}$$

where Λ is a constant representing the rate of change, i.e. either increase or decrease.

Table 3.1: Comparison of biological systems and energy systems.

Sl.No.	Lotka-Volterra model	Technology substitution model	Term
1.	Population of prey	No. of households using old technology	X_1
2.	Population of predator	No. of households using new technology	X_2
3.	Carrying capacity of prey	Maximum no. of households that can use old technology	N_1
4.	Carrying capacity of predator	Maximum no. of households that can use old technology	N_2
5.	Fractional growth rate of prey	Entry rate of old technology	k_1
6.	Fractional growth rate of predator	Entry rate of new technology	k_2
7.	Fractional death rate of prey	Exit rate of old technology	d_1
8.	Fractional death rate of predator	Exit rate of new technology	d_2
9.	Influence of predator on prey	Interaction factor of new technology on old one	b_1
10.	Influence of prey on prey	Interaction factor of old technology on new one	b_2
11.	Interaction medium - time	Interaction medium - income	I

This model predicts an exponential monotonic change.

$$X = A \exp(LI) \tag{3.17}$$

where A is an arbitrary constant.

This model may be improved upon by demanding that the system stabilises with income. This may be achieved by making the rate of change non-linear. This is equivalent to saying that Λ depends on X. Again, the dependence of Λ on X may be modelled by a linear function. For stability, it is apparent that Λ should decrease.

$$\Lambda = \mu - \alpha X \tag{3.18}$$

where μ and α are constants.

Incorporating this in equation (3.16), we have

$$\frac{dX}{dI} = X(\mu - \alpha X) \tag{3.19}$$

This leads to the solution

$$X = \frac{A\mu}{\alpha A + \exp(-\mu I)} \tag{3.20}$$

This model represents a system which starts at a particular state and exponentially stabilises to another state. This exponentially stabilising model may be used for modelling competing systems.

Generally, competing systems are modelled using a set of simultaneous equations. In other words, X_1 and X_2 are interdependent. Consider the influence of X_2 on X_1. Essentially, it is the rate of change that is altered. Once again, the idea of linear dependence of μ on X_2 is logical and is the simplest.

$$\mu = \nu - \beta' X_1 \tag{3.21}$$

where ß' is a parameter denoting the degree of influence of X_2 on X_1. Substituting equation (3.21) in (3.19), we have

$$\frac{dX_1}{dI} = X_1(\nu - \alpha X_1 - \beta' X_2) \quad (3.22)$$

Equation (3.22) may be written in an alternative form for easy interpretation.

$$\frac{dX_1}{dI} = K_1 X_1 (N_1 - X_1 - \beta X_x) \quad (3.23)$$

where

$\nu = K_1 N_1$

$\alpha = K_1$ and

$\beta = \beta' / K_1$

The variation of X_2 may be derived similarly and the resulting equation is

$$\frac{dX_2}{dI} = K_2 X_2 (N_2 - X_2 - \alpha X_1) \quad (3.24)$$

Interpretation of the system of coupled equations: Equations (3.23) and (3.24) represent the competition in terms of the rate of change of the number of households using one technology as being affected by the number of households using the other technology. This change is in addition to the inherent change due to its own parameters.

A special case of interest considered here is where the carrying capacities vary linearly with income I. In practical modelling, a polynomial variation of these with income I is appropriate. The coefficients α and ß are measures of the competition. If, $\alpha = 0$, the system T_1 is unaffected by T_2. In such cases, X_1 stabilises, as $I \to \infty$, to a value N_1 at a rate K_1. So K_1 encompasses the isolated rate of change. N_1 sets a limit on X_1. It may be termed as the carrying capacity of T_1. Similar arguments apply for T_2. The toughest competition is when $\alpha = \beta = 1$.

In the general case, the parameters N_1, N_2, α and β are dependent on I, which leads to interpreting them as varying, making the system of equations inhomogeneous.

Solution of the system in (3.23) and (3.24): The solution gives the variation of X_1 and X_2 with I subject to the initial state of the system given by $X_1(0) = X_1(I = 0)$ and $X_2(0) = X_2(I = 0)$.

In the general case, it is not possible to solve the system of non-linearly coupled equations analytically for all I. In special cases, a direct solution involving the ratio (X_2/X_1) may be attempted.

Consider a special case.

$\alpha = \beta = 1$, and $K_1 = K_2 = K$, independent of I.

$\alpha = \beta = 1$ indicates that the influence of a technology by the other is symmetric. $K_1 = K_2$ indicates the identical individual growth. Also, let N_1 and N_2 vary linearly with I, i.e. $N_1 = p_1 + q_1 I$ and $N_2 = p_2 + q_2 I$ where p_1, p_2, q_1, q_2 are constants.

$$\frac{dX_1}{dI} = KX_1(N_1 - X_1 - X_2) \tag{3.25}$$

or

$$\frac{1}{X_1}\left(dX_1/dI\right) = K(N_1 - X_1 - X_2) \tag{3.26}$$

and

$$\frac{dX_2}{dI} = KX_2(N_2 - X_2 - X_1) \tag{3.27}$$

and

$$\frac{1}{X_2}\left(dX_2/dI\right) = K(N_2 - X_2 - X_1) \tag{3.28}$$

Subtracting, we get

$$\frac{1}{X_2}\left({dX_2}/{dI}\right)\frac{1}{X_1}\left({dX_1}/{dI}\right) = K(N_2 - N_1) \tag{3.29}$$

or

$$\frac{d}{dI}(\ln X_2 - \ln X_1) = K(N_2 - N_1) \tag{3.30}$$

After substituting for N_1 and N_2 and simplifying, we get,

$$\frac{d}{dI}(\ln X_2 - \ln X_1) = K(p - qI) \tag{3.31}$$

where

$p = K(p_2 - p_1)$ and $q = K(q_2 - q_1)$.

Integrating (3.31) we get

$$\ln\frac{X_2}{X_1} = C + pI + qI^2 \tag{3.32}$$

where C is an arbitrary constant, depending on the initial state of the system, i.e. X_2/X_1 ($I = 0$). The constants C, p and q may be obtained by regression analysis of ln (X_2/X_1) with respect to I. These may be thought of as representing the given model.

As we can see, the model developed above bears a strong resemblance to the ecological models for predator-prey interactions and for the competition between species.

Conditions: What happens if the competition is equal/identical/tough. In the general case, the solution is

$$\frac{X_2}{X_1} = A\exp[C(I - I_0)] \tag{3.33}$$

where A is initial condition, i.e. X_2/X_1 at $I = I_0$ and

$C = (N_2 - d_2/k_2) - (N_1 - d_1/k_1)$

If the competition is identical, i.e. $N_1 = N_2$, $d_2 = d_1$, $k_2 = k_1$, then, the solution must differ only by the initial condition, i.e. the arbitrary constant A.

Let

$$\frac{X_2}{X_1} = A_1 \exp\left[C(I - I_0)\right] \tag{3.34}$$

Now, if we interchange X_1 and X_2 in the equation, we get

$$\frac{X_1}{X_2} = A_2 \exp\left[C(I - I_0)\right] \tag{3.35}$$

But, at $I = I_0$,

$$\frac{X_2(0)}{X_1(0)} = A_1 \; and \frac{X_1}{X_2} = A_2$$

or,

$$A_1 \times A_2 = 1$$

Multiplying (3.34) and (3.35)

$$\begin{aligned} 1 &= A_1 A_2 \exp\left[C\,(I-I_0)\right] \exp\left[C\,(I-I_0)\right] \\ &= \exp\left[C\,(I-I_0) + C\,(I-I_0)\right] \\ &= \exp\left[2\,C\,(I-I_0)\right] \end{aligned} \tag{3.36}$$

or,

$$\exp\left[2C\,(I-I_0)\right] = 1 \tag{3.37}$$

This is possible only if $C = 0$. This is seen even in (3.30)

Not identical: If the systems are not identical, $C \neq 0$. Since the solution varies exponentially, (i.e. $d_1 = d_2$, $k_1 = k_2$)

$$C = (N_1 - d_1/k_1) - (N_2 - d_2/k_2)$$

$$= (N_1 - N_2) - (d/k - d/k)$$

$$= (N_1 - N_2) \qquad (3.38)$$

Since $C \neq 0$, $N_1 - N_2 \neq 0$ or $N_1 \neq N_2$

Identical case or assuming the solution: Suppose, the identical conditions are also the same ($A_1 = A_2$) and let $N_1 \neq N_2$, i.e. $C_1 \neq C_2$

$$\frac{X_2}{X_1} = \exp\left[C_1 (I - I_0)\right] \qquad (3.39)$$

For equations with X_2 and X_2 interchanged, then

$$\frac{X_2}{X_1} = \exp\left[-C_2 (I - I_0)\right] \qquad (3.40)$$

From (3.38) and (3.39), we can get $C_1 = C_2$

$$\frac{X_2}{X_1} \rightarrow C_1 \text{ when } C_1 = (N_1 - N_2) \qquad (3.41)$$

$$\frac{X_2}{X_1} \rightarrow C_2 \text{ when } C_2 = (N_2 - N_1) \qquad (3.42)$$

i.e. $C_1 = C_2$ and $N_1 = N_2$

Extension and generalisation: The present model can be applied even when more than two competing technologies exist. The model may be generalised to a system consisting of an arbitrarily finite number of competing technologies S_i, $i = 1,2.....M$, represented by E_i, the medium of interaction being Z.

$$\frac{dE_i}{dZ} = K_i\left(N_i - E_i - \sum_{j=1} \beta_{ij} E_j\right) E_i \qquad (3.43)$$

This model approximates the general situation under the assumption that every given technology undergoes four different phases: learning, growth, saturation and decline.

When a new technology competes and replaces fully or partially the old technology, the new technology is never an exact equivalent of the old. Hence, it is not proper to compare the economics of these two through the unit prices. For example, in the case of competition between firewood and kerosene, it is not valid to compare the average price of a kerosene stove with that of a wood stove. In the first place, a firewood stove has no initial cost, transportation cost, etc. However, the efficiency, unbreakability and smokelessness of a kerosene stove permit the customer to opt for it. Thus, the alternative technologies do not compete on the basis of prices alone but rather on the basis of the maximum utility provided per unit cost.

The following model explores the economic factors underlying the shifting of households from one energy-utilising device to another.

Consumer Discount Rates and Energy Carrier Choices

The competition model developed above has explained the income-dependence of the ratio (X_1/X_2) of households utilising the devices D_1 and D_2 as well the second-degree polynomial shape of the (X_1/X_2) vs I curves. It is now necessary to explore the economic factors underlying the shifting of households from one energy-utilizing device D_2 to another D_2.

The choice of an energy carrier by a household depends mostly on the capital cost of the device (stove). The efficiencies with which the carriers are utilised (efficiency of the device) vary, as also the capital costs. Since the capital cost of an efficient device is high, low-income households cannot afford to invest in it. Therefore, even if the return on investment is high, these households will not be able to invest in efficient devices. Thus, we can say that there is a relationship between the household income and the consumer discount rate.

In countries like India, households do not know their discount rates due to a lack of knowledge regarding energy carriers and their utilising devices (Ruderman *et al.*, 1987). Consumer discount rates are not easy to come by and, therefore, a significant contribution of the present study is the extraction of discount rates. Using the life-cycle cost analysis, and the return on investment principles, the consumer discount rates can be obtained under the assumption that the consumer shifts from one carrier to another if his internal rate of return is greater than the consumer discount rate. Using the discount rates, the level of conservation equipment can be predicted.

Analysis of consumer discount rates

The analysis assumes that consumer behaviour is characterised by the capital cost of the energy-utilising device. Even if the consumer does not decide on the basis of the life-cycle cost (LCC), the discount rates can be used as a measure of market imperfection. This analysis examines different energy carriers (in terms of devices) used by different households and calculates the LCC taking into consideration the capital cost of the device and its life, operating cost, efficiency of the device, energy carrier price, etc. In calculating the LCC it was assumed that the consumers do not anticipate any escalation in the prices of energy carriers.

The capital cost of the device increases with increasing efficiency, whereas the operating costs are more for the less efficient devices. Hence a large investment in an efficient device would result in future savings. By determining how some of the carriers (in terms of efficient devices) must pay back the incremental investment, one can calculate the internal rate of return (IRR). The shorter the payback period, the larger the IRR[1].

Calculation of internal rate of return

If the consumer is economically rational, he will distinguish between efficient and inefficient devices if the net present value of energy savings equals the additional investment on the efficient device. We can use this observation to estimate the consumer discount rates using the life-cycle cost (the sum of the present values of all the costs over the lifetime) analysis.

[1] This is subject to the limitation of initial investment capability, convenience of use and availability.

The life cycle cost of the device D_1 is calculated as:

$$LCC\,(1) = I(1) + I(1)\,PV\{1A, d_{eff}, (N-1)\} + E(1)\,PV\{1A, d, L(2)\} \quad (3.44)$$

where

$LCC\,(1)$ = Life-cycle cost of the inefficient device D_1 (Rs)

$L\,(1)$ = lifetime (years)

$I\,(1)$ = Initial cost (Rs)

$E\,(1)$ = Annual expenditure on energy (Rs)

$PV\{1A, d_{eff}, (N-1)\}$ = present value of a stream of $(N-1)$ Rs 1 payments.

Let the corresponding values for a more efficient device D_2 with a lifetime $L(2)$ be $I\,(2)$ and $E(2)$.

The $I\,(1)$ appears in the term $[I\,(1)]\,PV\{1A, d_{eff}, (N-1)\}$ to take care of the cost of replacing the shorter life device D_1 $(N-1)$ times so that the duration of operation coincides with the life of the longer lifetime device D_2.

The corresponding life-cycle cost for the device D_2 is

$$LCC(2) = I\,(2) + E(2)\,PV\{1A, d, L(2)\} \quad (3.45)$$

If a household shifts from device D_1 to D_2, the net present value (NPV) for the shift is given by

$$\begin{aligned} \text{NPV} &= LCC(1) - LCC(2) \\ &= [I(2)-I(1)]+I(1)PV\{1A, d_{eff}, (N-1)\}+[E(1)-E(2)]\,PV\{1A, d_{eff}, N\} \\ &= \underline{\Delta} I + I(1)\,PV\{1A, d_{eff}, (N-1)\} + \underline{\Delta} E\,PV\{1A, d_{eff}, N\} \quad (3.46) \end{aligned}$$

The internal rate of return IRR ($D_1 \rightarrow D_2$) for the shift is obtained by setting the NPV = 0 and solving for d_{eff}. In other words, the IRR ($D_1 \rightarrow D_2$) can be obtained by solving the equation $LCC(1) = LCC(2)$ for d_{eff}.

This IRR is based on the assumption that the consumer is offered several investment options and he prefers the one with the lowest investment (Hausman, 1979).

Since the LCC includes capital cost, operation and maintenance costs, it is easy to compare the costs of appliances whose performance characteristics are similar.

Though the value of IRR ($D_1 \rightarrow D_2$) reveals the benefits to a household by shifting from the device D_1 to the device D_2, it does not indicate whether the household decides that this internal rate of return is large enough to undertake the shift. Such decisions are based upon the consumer discount rate (CDR) used by the household and, in particular, upon how the IRR ($D_1 \rightarrow D_2$) compares with its CDR. The decision rule is as follows: if IRR ($D_1 \rightarrow D_2$) < CDR, then the switch $D_1 \rightarrow D_2$ is rejected, but on the other hand, if IRR ($D_1 \rightarrow D_2$) > CDR, then the switch is favoured.

Estimation of consumer discount rates

The consumer discount rate of a household depends upon its capital resources which determine whether the household can afford to invest an amount of capital ($\wedge Io$) now on a more energy-efficient device in order to reap regular benefits later in the form the energy savings (Dubin and McFadden, 1984). This means whether the household has the capacity to sacrifice scarce capital today in order to derive benefits in the future. If this preparedness to defer current expenditures depends, in general, upon its income, and in particular, upon its disposable income, then it is obvious that the consumer discount rate must depend upon the income. If the income is high, then the household has the ability to make current investments in order to derive future benefits, and, therefore, the consumer discount rate is small. Assuming that this consumer discount rate decreases exponentially with income, one can write

$$\text{CDR} = A \exp(-BI) \text{ or } \ln \text{CDR} = \ln A - BI \tag{3.47}$$

where A and B are constants and I is the household income.

It follows from this equation that as the income of households and, therefore, their capital resource increases, their consumer discount rates decrease, and the households are willing to invest larger and larger amounts of capital ($\wedge Io$) on costlier devices in order to reap regular future benefits from more energy-efficient and longer-life devices. This means that as the household income increases, there will be a series of device shifts, $D_1 \rightarrow D_2$, $D_2 \rightarrow D_3$, $D_3 \rightarrow D_4$, etc.

The models developed above are typical of an urban energy system in a developing country like India. We must apply these models to a representative urban setting. We consider the city of Bangalore in India which is a rapidly urbanising centre so that the models developed here can be effectively validated.

4

TRADITIONAL FUELS IN AN URBAN SOCIETY

The forest is a peculiar organism of unlimited kindness and benevolence that makes no demands for its sustenance and extends generously the products of its life activity; it affords protection to all beings, offering shade even to the axeman who destroys it.

– Gautam Buddha

The importance of traditional energy in general, and firewood and charcoal in particular, in the national energy system was realised fairly early. India has been a pioneer in this matter. As early as 1974, the Report of the Fuel Policy Committee (of the Planning Commission, 1974) stated: *The most important fact about the energy situation in India is that nearly half of the total energy consumed in the country comes from non-commercial sources*. Despite this, the details of firewood consumption, for example, are not as well understood as the consumption of a commercial energy carrier like electricity. Fortunately, during the past two decades, a number of rural energy studies (NCAER, 1978; Pachauri, 1978; Reddy, *et al*., 1977; ASTRA, 1982) have been conducted which threw light on the pattern of non-commercial energy usage in rural areas. But a few comparable studies exist for urban settlements. There are, however, some notable exceptions (NCAER, 1958; Sumithra, 1981). Since non-renewable sources of energy like petroleum are limited and hence expensive, the vast potential of wood as an energy source merits serious and immediate consideration. This section tries to focus attention on the traditional sources of energy, namely, firewood, charcoal and coal.

Wood is the fourth largest energy source in the world and the first in India. The study of wood energy is of particular importance for India in view of the total dependence of most poor people on this energy resource. Also, the high rate of population growth, rising prices of commercial energy carriers, viz. electricity, gas and petroleum products, suggest that wood energy consumption will increase in future. However, wood energy suffers from problems of accessibility (forests, hydro-electric power) and low flexibility in usage, although it can be processed into many forms. Every year plants store ten times the present annual consumption of energy. This large quantity of energy cannot only contribute to the country's energy resources, but can effectively provide a wide range of feedstock for the organic chemical industry, thus conserving precious non-renewable petroleum products. Moreover, it prevents ecological imbalance.

Firewood and charcoal consumption often produce a negative environmental impact because they can be associated with deforestation. At first, it was assumed that deforestation takes place because villagers depend on firewood for cooking fuel. However, many studies (ASTRA, 1982; Sataye and Meyers, 1985) have shown that, in several rural areas, this need not be the case. The point is that villagers rely predominantly on twigs, roots and thin branches, gathered mainly by women and children. It has, therefore, been suggested that, apart from rural brick-burning which requires logs from felled trees, deforestation must be attributed mainly to urban/industrial consumption of firewood and charcoal. But, this suggestion needs substantiation from studies of firewood and charcoal usage in urban concentrations. This study tries to fulfil that need.

Firewood

The supply

The species of firewood that were being used in Bangalore are *casuarina, haldu, bijasal, laurel, irul, pongamia, tamarind, coffee, persian lilac* and *kindal*. Between 1981 and 1990 there was a significant change in the species-mix and their share. *Casuarina* constituted 30 per cent of the total consumption in 1981, but, in 1990, its consumption came down to only 22 per cent. The share of *haldu, bijasal, laurel* and *irul* increased from 15 to 30 per cent, while that of *pongamia* and *tamarind* decreased from 25 to 18 per cent. *Coffee* and *eucalyptus* were no longer being used. During 1990, retailers reported that *eucalyptus* was mainly used as lumber (in the form of long poles) rather than as firewood perhaps because it is at least 25 per cent costlier than other species. The share of *babul, axlewood* and *banyan* increased from 15 to 20 per cent. *Vilayate babool* which was not used in 1981 was being used in 1990 with a 5 per cent share of the total consumption. *Kindal* and *persian lilac* account for the remaining 5 per cent.

The inflow of firewood to Bangalore city comes from various sources. Of the total, about 50 per cent was from the forests and privately owned plantations near Chintamani, Srinivasapura, Madanapalle and Chittoor about 100-150 km away from Bangalore. The main species coming from these forests were found to be *tamarind, babul, axlewood, pongamia* and *kindal*.

About 35 per cent of the total firewood inflow to Bangalore came from private lands in Nelamangala, Anekal, Magadi, Malur, Ramanagaram, Kumbalgud and Bidadi, i.e. from within a radius of 30-40 km. *Casuarina, tamarind* and *banyan* were the main species coming from these sources. The government forests in Varahi, Dharwad, Sirsi, Shimoga, Haliyal, Sagar, Mercara, Hunsur and Kankinakota supplied about 6-7 per cent. These sources are at distances of 300-400 km from Bangalore, and the supply was mainly in the form of *haldu, bijasal, laurel, kindal* and *pongamia*. The remaining 8-9 per cent of the total firewood supply came from government forests in Goa. The sources of supply in this category were Collam, Kalay, Sanvardenkachoram and Londa which are at distances of 650-700 km from Bangalore. The main species supplied by these forests are *bijasal, haldu, laurel, kindal* and *pongamia*.

The Forest Department of the Government of Karnataka and private contractors also supplied firewood to Bangalore. The Forest Department supplied firewood to domestic consumers through cooperative societies which receive monthly allotments that were then sold to ration-card holders. Private contractors obtained their supply of firewood from two sources, viz. (i) private lands (including their own) and (ii) government auctions. These contractors then sold the firewood to retailers through agents. Table 4.1 gives the quantities of firewood supplied during the years 1980-90 by private contractors and the Forest Department. The data show that the average supply

Table 4.1: Quantity of firewood supplied to Bangalore (1980-90).

Year	Supply from various sources (tonnes)		
	Forest Department	Private contractors	Total
1980-81	21035	393174	414209
81-82	22392	382887	405279
82-83	20960	398371	419331
83-84	20734	411406	432140
84-85	19205	.419021	438226
85-86	19374	427163	446537
86-87	17540	446824	464364
87-88	19768	454783	474551
88-89	na	461178	461178
1989-90	na	471617	471617

na = not available

from the private contractors in any year is approximately 20 times the corresponding quantity supplied by the Forest Department.

Transportation

The two principal modes of transporting firewood to the city were railway wagons and trucks, but a small quantity of firewood also came in by means of the bullock-cart. In addition, observation showed that some residents of the city cut firewood from roadside trees outside the city and brought it in as head loads. The bulk (85 per cent) of the firewood was transported to Bangalore through trucks – and this came predominantly from private lands. Of the total inflow, about 10 per cent came via rail, mainly from government forests in North Kanara and Goa. Table 4.2 gives the number of railway wagons and trucks that brought firewood to Bangalore during the years 1980-90.

The data in Table 4.2 indicate that the number of trucks (with a capacity of 6-8 tonnes per truck) that brought firewood is nearly ten times more than that of railway wagons (with a capacity of 11.5-12.5 tonnes/wagon). From 1980 to 1990, the data regarding the number of trucks that transported firewood indicated an increase, within a 99 per cent confidence level, while the railway wagons registered a decrease within a 90 per cent confidence level. The smaller coefficient of variation in the case of trucks indicates that, with regard to firewood, they are a steadier transport mode than railways. In addition, about 100 bullock carts per day transported firewood into the city, and a few

Table 4.2: Number of trucks and railway wagons that transported firewood (1980-90).

Year	Trucks (no.)	Wagons (no.)
1980-81	42087	3780
81-82	41144	3650
82-83	42037	3565
83-84	43679	3399
84-85	47260	3301
85-86	47948	3290
86-87	48608	3385
87-88	49238	3165
88-89	50236	3093
89-90	50992	2905

tonnes also arrived by headloads. The magnitudes of inflow can be estimated by considering each transportation mode in turn and then obtaining the total.

(a) *Trucks*: The quantity of firewood brought into the city by trucks was obtained from observations at the five weigh-bridges located at the entry roads to Bangalore. There were two types of weigh-bridges with capacities of 16 tonnes and 20 tonnes, respectively. The average tare or empty weight of a truck was found to be 5.52 ± 0.65 tonnes from a sample of 100 empty lorries. From the 20-tonne weigh-bridge, the weight of a firewood-loaded truck was found to be 17.15 ± 1.36 tonnes from a 100-truck sample. This corresponds to an average firewood load of 11.64 tonnes per truck. Similarly, at the 16-tonne weigh-bridge, the weight of a truck loaded with firewood was determined to be 13.66 ± 1.65 tonnes from a sample of 100 trucks. This works out to an average load of 8.14 tonnes per truck. From this data and from the number of firewood-loaded trucks coming in via the two types of weigh-bridges, the inflow of firewood through trucks was calculated.

(b) *Wagons*: Wagons of both the Southern Railway and the South Central Railway were used to transport firewood to the city. Based on a 100-wagon sample from the Southern Railway, it was found that the quantity of firewood transported was 11.9 ± 1.3 tonnes/wagon. In the case of the South Central Railway, their charges were on the basis of 11.5 tonnes/wagon. From this data and the number of wagons used for transporting firewood, the quantity of firewood brought into Bangalore by rail was estimated.

Table 4.3 presents the data on the quantity of firewood transported through railway wagons and trucks.

The figures in the table indicate that during the 10 years from 1980-90, the railways used, on an average, 3,442 wagons/year to bring 41,304 tonnes of firewood into Bangalore. This works out to 111.5 tonnes/day carried in approximately 10 wagons. The total quantity of firewood transported through trucks and railway wagons per day ranged from 1,101 tonnes in 1980-81 to 1,320 tonnes during 1989-90.

The results of a sample survey showed that a certain quantity of firewood per day was brought into the city through bullock-carts and headloads. On an average, about 100 bullock-carts carried firewood to Bangalore every day, each carrying an amount of 500 kg. Hence, the inflow of firewood via bullock-carts was estimated to be 50 tonnes per day. The outskirts of the city were surveyed and it was found that about 300 people brought headloads of firewood into the city daily. A sample of 20 headloads was weighed and it was observed that each headload weighed approximately 25-30 kg. Hence, the total quantity of firewood flowing into the city via headloads was estimated to be nearly 10 tonnes per day.

Distribution

The distribution of firewood in Bangalore was handled by cooperative societies and commission agents, with the former obtaining their supplies from the Forest Department and the latter from private lands and auctions conducted by the Forest Department. The commission agents in turn distributed their firewood to retail depots and to direct

Table 4.3: Transportation of firewood through trucks and raiways.

Year	Quantity of firewood transported (tonnes)		
	Trucks	Railways	Total
1980-81	369276	43470	412746
81-82	360081	41975	402056
82-83	378333	40998	419331
83-84	393111	39089	432200
84-85	416801	37962	454763
85-86	422832	39477	462309
86-87	425205	40447	465652
87-88	439519	37990	477499
88-89	450619	38980	488599
89-90	457754	36187	493941

purchasers (Figure 4.1).

There were 23 cooperative societies in Bangalore which handled firewood. In addition, there were 700 registered firewood retail depots, and an approximately equal number of unregistered depots. The consumers obtained the bulk (85 per cent) of their firewood requirement from retail depots. Direct purchases from agents and procurement from cooperative societies accounted for 10 per cent and 5 per cent respectively. The total quantity of firewood handled by all these channels was estimated to be 1,236 tonnes/day.

Consumption

Bangalore's firewood consumers fall into 11 categories: (1) households, (2) dyeing factories, (3) bakeries, (4) restaurants, (5) soap factories, (6) choultries, (7) hostels, (8) cremation grounds, (9) canteens, (10) road asphalting agencies and (11) industries.

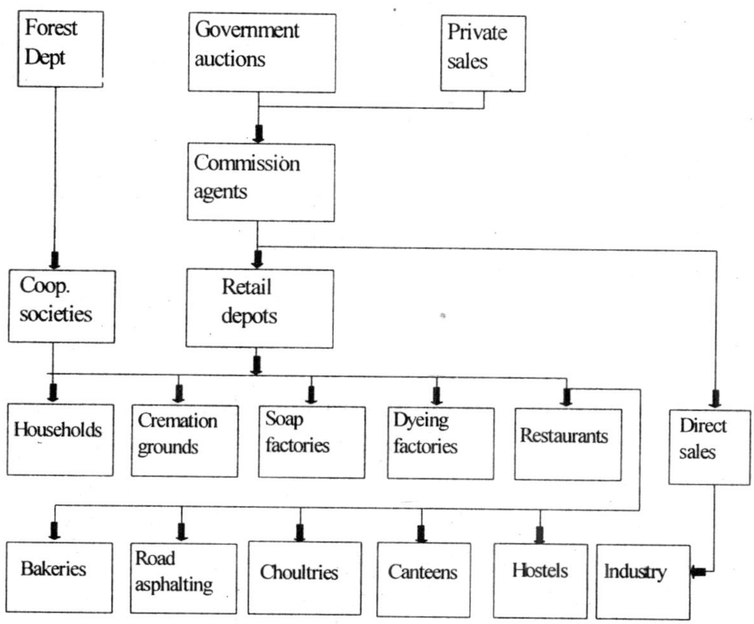

Figure 4. 1: Distribution of firewood to different consumers.

These can be further divided into the following categories: (i) residential, (ii) commercial, and (iii) industrial.

Residential sector: A household sample survey showed that only 16.6 per cent of the sample households used firewood for both cooking and water-heating, 18.8 per cent used it only for water-heating and 64.6 per cent did not use any firewood at all either for cooking or for water-heating. Depending on their profession, the households were divided into various categories.

Considering casual labourers as a special category, 80.9 per cent of these households depend wholly on firewood for both cooking and water-heating and only 8.2 per cent on other fuels which, in their case, means dung-cakes, twigs and biomass pickings. In complete contrast was the category of households in service, business and professions; 73 per cent of the households in this category used fuels other than firewood, such as electricity, kerosene and LPG, for both cooking and water-heating, and only 27 per cent used firewood for both the tasks (Table 4.4).

The type of fuel chosen for cooking and water-heating depends, as one would expect, on the household income. The income-dependence of firewood usage is confirmed by the fact that, out of the 166 households in the sample that rely wholly on firewood, 18.1 per cent had an income of up to Rs 250 per month and another 42.2 per cent households had a monthly income range of Rs 251-500, that is, 60.3 per cent of the households which used firewood for both cooking and water-heating had a monthly income of Rs 500 or less and only 9 per cent of the category of households totally dependent on firewood had a monthly income of Rs 1,000 or more (Table 4.5).

Table 4.4: Distribution of households by profession using various energy carriers.

Profession of the head of the household	Households using particular fuels			
	Firewood for cooking and water-heating	Firewood for water-heating only	Fuels other than firewood	Total
Casual labour	89 (80.9)	12 (10.9)	9 (8.2)	110 (100.0)
Skilled labour	41 (17.7)	37 (15.9)	154 (66.4)	232 (100.0)
Service (clerical and allied)	13 (7.2)	52 (28.9)	115 (63.9)	180 (100.0)
Service (higher level)	2 (2.3)	15 (17.4)	69 (80.2)	86 (100.0)
Business (small)	15 (16.3)	27 (29.3)	50 (54.4)	92 (100.0)
Business (big)	-	3 (7.1)	39 (92.9)	42 (100.0)
Professional	-	10 (9.3)	98 (90.7)	108 (100.0)
Others	6 (4.0)	32 (21.3)	112 (74.7)	150 (100.0)
Total	166 (16.6)	188 (18.8)	646 (64.6)	1000 (100.0)

Note: Numbers in parentheses represent percentages.

Table 4.5: Distribution of households according to family income and energy carrier usage.

Household income (Rs/month)	Households using particular energy carriers			
	Cooking firewood Water-heating firewood	Cooking – other carriers, Water-heating - firewood	Other fuels	Total
<250	30 (57.8)	10 (19.2)	12 (24.0)	52 (100.0)
251-500	70 (45.8)	43 (28.1)	40 (26.1)	153 (100.0)
501-1000	51 (16.0)	73 (22.9)	195 (61.1)	319 (100.0)
>1000	15 (4.2)	62 (14.0)	399 (84.8)	476 (100.0)
Total	166 (16.6)	188 (18.8)	646 (64.6)	1000 (100.0)

Note : Numbers in parentheses represent percentage.

The household income was divided by the family size and the per capita monthly income (PCMI) was obtained. The energy carriers used by various PCMI groups are shown in Table 4.6.

It can be concluded from Table 4.6 that households in the per capita monthly income groups I, II and III constituted 14.8 per cent, 17 per cent and 11 per cent, respectively, of the sample, but they accounted for 42.2 per cent, 30.1 per cent and 12.1 per cent, respectively, of the households using firewood for both cooking and water-heating. Alternatively stated, households with per capita monthly incomes of less than Rs 150 constituted 31.8 per cent of the survey sample, but accounted for 72.5 per cent of the households which depend on firewood for both cooking and water-heating. The situation is even clearer with households having a per capita income of less than Rs 200. They constituted 42.8 per cent of the sample, but almost 85 per cent of the households using firewood for cooking and water-heating belonged to this income group.

Table 4.6: Distribution of households according to PCMI and firewood usage.

PCMI (Rs)		Households using FW		Not using FW	Total
Group	Range	Cooking and Water-heating	Water-heating only		
I	<100	70 (47.3)	33 (22.3)	45 (30.4)	148 (100.0)
II	100-149	50 (29.4)	34 (20.0)	86 (50.6)	170 (100.0)
III	150-199	20 (18.2)	31(28.2)	49(53.6)	110 (100.0)
IV	200-249	15 (12.4)	36 (29.8)	70 (57.8)	121 (100.0)
V	250-299	7 (5.8)	32 (26.4)	82 (67.8)	121 (100.0)
VI	300-349	3 (2.2)	17 (12.3)	118 (85.5)	138 (100.0)
VII	>350	1 (0.5)	5 (2.6)	186 (96.9)	192 (100.0)
	Total	166 (16.6)	188(18.8)	646(64.6)	1000(100.0)

Note: Numbers in parentheses represent percentages.

To reveal the income dependence of firewood consumption the total firewood dependence index (TFDI) can be used. The TFDI values for different income categories are given in Table 4.7.

Table 4.7: Firewood dependence indices of households.

PCMI (Rs)		Mid-point of income range (1)	% of households in the category (2)	% of households using FW for cooking and water-heating (3)	TFDI(4)= 3/2
Group	Range				
I	<100	50	14.8	42.2	2.85
II	100-149	125	17.0	30.1	1.77
III	150-199	175	11.0	12.1	1.10
IV	200-249	225	12.1	9.0	0.74
V	250-299	275	12.0	4.2	0.35
VI	300-349	325	14.8	1.8	0.13
VII	> 350	450	19.3	0.6	0.03
	Total		100.0	100.0	

It is seen that a per capita monthly income of Rs 200 is the benchmark. The category having an income of below Rs.200 bears a greater burden of total firewood dependence than the category's proportion in the sample, and the categories having an income above Rs 200 had much less dependence than their percentage in the sample. In fact, if the mid-point of the income range is chosen as an independent variable and the total firewood dependence index as the dependent variable, the survey observations show a linear correlation:

$$\text{TFDI} = 4.28 - \underset{(0.006)}{0.01}\, X \text{(Mid point of income range)} \qquad (4.\ 1)$$

$R^2 = -0.84.$

(*Note*: Figure in parenthesis represents the standard error.)

The result indicates that with each unit increase in the mid-point of income the TFDI decreases by 0.01 units. Secondly, the Lorenz curve, i.e. the plot of the cumulative percentage of households using firewood *vs* the cumulative percentage of households in the various income ranges, is highly skewed (Figure 4.2) showing a marked inequality in the burden of dependence on firewood. As expected, the lower-income groups had a far greater dependence on firewood.

The firewood consumption results described above show that there is a variation in the consumption of the individual households. This may be due to a variation in the characteristics of the individual households. At the time of the survey, it was observed

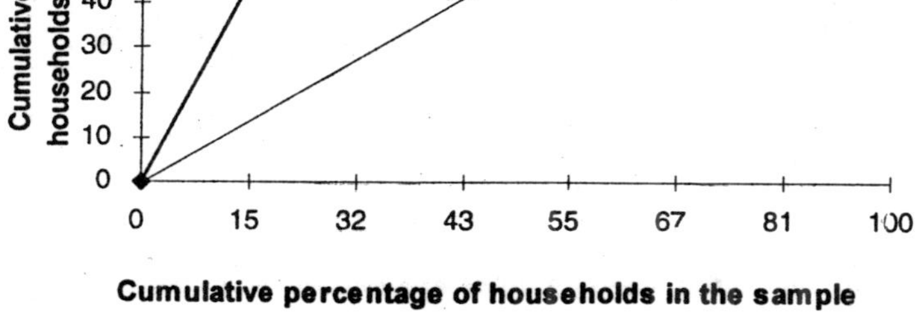

Figure 4.2: Lorenz curve for firewood usage.

that the household firewood consumption (FWC) varied with family income (FI), and with family size (FS). It was also observed that many of the households used more than one fuel. Charcoal (CH), kėrosene (KR) and waste, including dung and agricultural waste (DG), are used as auxiliary fuels and, therefore, have an influence on the consumption of firewood. In view of this expected relationship, household firewood consumption was regressed on family income, family size and the consumption of charcoal, kerosene and other waste products, using the multiple regression technique. The results are shown in the following equation (Figure 4.2).

$$FWC = \underset{}{32.550} + \underset{(1.65)}{1.416\, FS} + \underset{(0.005)}{0.036\, FI} - \underset{(0.406)}{2.813\, CH} - \underset{(0.63)}{1.99\, KR} - \underset{(0.126)}{0.82\, DG} \quad (4.2)$$

$$R^2 = 0.70$$

(*Note*: Figures in parentheses represent standard errors.)

Each regression coefficient indicates the marginal rate of change of the dependent variable caused by the independent variable, i.e. it is the partial differential of the dependent variable with respect to the explanatory variable. The results show that by increasing the family income by one unit, the firewood consumption increases by 0.036 units and also by increasing the family size by one unit, the consumption increases by 21.416 units. Hence, it is the family size rather than the family income which determines the amount of firewood used. Further, with an increase in the consumption of auxiliary fuels, firewood consumption decreases (as shown by the negative signs of the coefficients) because these fuels are used as substitutes for firewood. Stepwise regression results show that the consumption of waste plays an important role in explaining the variation in the firewood consumption. The high values of t in the table indicate that the regression coefficients are significant (at the 90 per cent confidence level). (Table 4.8)

Table 4.8: Regression results for the firewood consumption.

Variable	Degrees of freedom	Mean (kg/m)	Std. dvn	Variable	Coefficient	Std.dvn	t value	R^2 (individual)	R^2 (step-wise)
Family size (no)	165	5.65	2.44	FS	21.460	1.650	13.01	0.51	0.51(FS only)
Family income (Rs.)	164	678.00	504.00	FI	0.036	0.005	7.20	0.16	0.59(FS+FI)
Charcoal consumption (kg/m)	163	2.66	6.88	CH	-2.813	0.460	6.12	0.02	0.64(FS+FI+CH)
Kerosene consumption (l/m)	162	3.70	3.60	KR	-1.990	0.630	3.16	0.04	0.67(FS+FI+CH+KR)
Dung consumption (kg/m)	161	15.10	21.90	DG	-0.820	0.126	6.51	0.002	0.70(FS+FI+CH+KR +DG)

Since the family size seems to be a major determinant of firewood consumption (R^2 = 0.51), it is of interest to see how the family size affects the per capita firewood consumption, i.e. to find the relation between the per capita firewood consumption (PCFWC) and the category of the family size of the sample households.

Taking 166 households which use firewood for both cooking and water-heating and classifying them into different family-size categories (FSC), the average monthly per capita firewood consumption (MPCFWC) in kg/cap/month in each category was regressed against the number of persons in the household, i.e. household size in that category. The regression result was as follows:

$$\text{MPCFWC} = \underset{}{33.20} - \underset{(0.07)}{0.84}\ \text{FS} \qquad (4.3)$$

$R^2 = 0.73$

(*Note*: Figure in parenthesis represents standard error.)

This interesting observation that the per capita firewood consumption decreases as the number of persons in the household increases is most probably due to the economies of scale.

The quantity of firewood consumed and the expenditure on firewood are also of interest. Table 4.9 presents the data for households which use firewood for both cooking and water-heating. It is seen that the average annual per capita firewood consumption for these households in each income group increases with increasing per capita income. It is not clear, however, whether this increase is due to greater food consumption or due to higher fractions of home-cooked food. In Income Group I, the annual per capita consumption is 331.6 kg which increases to 456 kg in group VII, an increase of 27.3 per cent. Using the January, 1987 firewood price of Rs 0.5/kg, the average expenditure on firewood is Rs 13.8 in the lowest income group and Rs19 in the highest income group. Further, the per capita monthly expenditure on firewood is a decreasing percentage of the per capita monthly income, constituting an estimated expenditure on firewood of 19.6 per cent of the total income in the case of the lowest income group. For water-heating also, it appears that, as the per capita income increases, the per capita consumption of firewood for water-heating increases in absolute magnitude. This result may be due to an increase in the frequency of hot water baths with increasing income. The per capita firewood consumption in Income Group I is 168.4 kg/year corresponding

Table 4.9: Consumption and expenditure on firewood used for cooking and water-heating only.

PCMI (Rs)		Per capita consumption (kg/year)		Expenditure as % of income	
Group	Range	Cooking and water-heating	Water heating only	Cooking and water-heating	Water-heating only
I	<100	332	168	20.0	10.0
II	100-149	366	191	12.9	6.4
III	150-199	377	193	9.5	4.7
IV	200-249	392	198	7.4	3.7
V	250-299	401	214	6.2	3.2
VI	300-349	404	228	5.1	3.0
VII	>350	456	245	3.8	2.0
	Average	390	211	9.2	3.9

to an expenditure of Rs 7/cap/month. In the case of Income Group VII, the consumption is 245.3 kg/cap/year with a monthly expenditure of Rs 10.20. In the case of these households, the per capita monthly expenditure is also a decreasing percentage of the per capita monthly income. The estimated expenditure on firewood for water heating is 10 per cent of the total income in the case of Group I, and 2.3 per cent in the case of Group VII.

The data contained in Table 4.9 can be used to estimate the amounts of firewood used for cooking and for water-heating. Assuming that the same amount of firewood consumed in households that use firewood only for water-heating is used for this purpose in households that use firewood for both cooking and water-heating, the shares of cooking and water-heating were estimated to be 52 per cent and 48 per cent, respectively. Table 4.9 shows that apart from the variation with per capita income, the average amount of firewood consumed for cooking works out to be 202.8 kg/capita/year. This quantity is low compared to the corresponding figure in rural areas, e.g., an average of 520.4 kg/capita/year in six specific villages of Kunigal taluk, Tumkur district, Karnataka (ASTRA,1982). Thus, the per capita firewood consumption for cooking in Bangalore is only 39 per cent of the corresponding consumption in villages. It has not been possible, however, to establish whether this reduced consumption is due to the use of dry logs and metal vessels in Bangalore in contrast to moist twigs and mud-pots in villages, or due to a greater frequency of eating out in restaurants in urban areas.

The extrapolation of the results obtained from the 1,000 household sample to the city of Bangalore has been carried out in the following way. The percentages of households in any income range using firewood for both cooking and water-heating and for water-heating only are assumed to be the same for the whole city as in the sample. Further, the per capita firewood consumption norms obtained in the sample are assumed to hold good for Bangalore. However, using these arguments, the total domestic consumption of firewood has been estimated to be about 995 tonnes per day (Table 4.10).

Commercial sector: There were about 1,850 bakeries in Bangalore that used firewood for baking bread, biscuits, etc. The firewood consumption norm was 0.6 kg of firewood for baking 1 kg of maida (i.e. wheat flour). The Food and Civil Supplies Department issued, on an average, about 60 tonnes of maida/day to the 1,150 bakeries that were registered. Hence, firewood consumption of the registered bakeries can be taken to be 35 tonnes/day. There were, in addition, about 700 unregistered bakeries which, according to the Bakeries Association, consumed about 1,000 tonnes of maida/month corresponding to a firewood consumption of about 20 tonnes/day. Thus, the total consumption of firewood was about 55 tonnes/day.

There were 1,869 restaurants in Bangalore of which 659 were of A and B grades, and 1,210 were of C and D grades. The A and B grades did not use firewood at all. As for the C and D grade restaurants, a survey carried out on a sample of 60 restaurants showed that about 50 per cent of them used firewood and that the average consumption

Table 4.10: Estimated firewood consumption in various income groups.

PCMI (Rs)		FW consumption (tonnes/day)					
Group	**Range**	**Households in the range (%)**	**Cooking and water-heating**	**Water-heating only**	**Total**	**% of total**	**Cumulative %**
I	<100	31.9	408.2	99.4	507.	51.0	51.0
II	100-149	12.7	101.4	39.4	140.	14.2	65.2
III	150-199	15.6	74.1	64.9	139.	14.0	79.2
IV	200-249	17.6	57.6	74.8	132.	13.3	92.5
V	250-299	8.6	13.2	34.6	47.	4.8	97.3
VI	300-349	10.0	6.0	19.4	25.	2.5	99.8
VII	>350	3.6	0.5	1.5	2.	0.2	100.0
	Total	100.0	661.0	334.0	995.	100.0	

was about 80 kg/day. Thus, the total daily consumption of firewood can be taken as 48 tonnes/day.

Industrial consumption: The main category of industries in Bangalore that used firewood for their boilers/furnaces were: (1) plywood factories, (2) brick and tile factories, (3) tea factories and (4) small-scale industries. Of the two plywood factories in Bangalore, one used firewood for its boilers. The consumption norm was 0.25 tonne of firewood for drying and hot-pressing 1,000 sq.ft of 1 mm thick plywood. This factory produced about 100,000 sq ft of plywood per day and half this quantity was dried and pressed with the heat energy from the combustion of firewood. Hence, the daily consumption was about 12.5 tonnes of firewood/day. There were four brick and tile factories that used firewood for burning bricks and tiles. They required 2 to 3 tonnes of firewood to manufacture 1,000 tiles. The total production of these factories was 20 million tiles/year of which one-fourth was burnt with firewood. This corresponds to a firewood consumption of about 9 tonnes/day. One tea factory used about 8 tonnes of firewood/day for curing tea leaves with a consumption norm of 1.3 kg firewood/kg tea leaves. There were also several small industries that utilised firewood for their furnaces and boilers. It was estimated that the daily consumption of these factories was about 18 tonnes/day. Thus, the total consumption of firewood by industries was taken as approximately 48 tonnes/day.

Dyeing and soap factories also use firewood. Except for a lone unit, all 176 dyeing factories utilised firewood as fuel. The consumption norm was 8.33 tonnes firewood/tonne of yarn that is dyed, and the total yarn dyed was about 7.5 tonnes/day (excluding the factory that did not use firewood). Hence, the daily consumption of firewood for all the dyeing factories was 62 tonnes/day. There were about 50 soap factories in Bangalore, approximately half of which used firewood for soap production. The total production of all the soap factories was about 250 tonnes of soap/month and the consumption norm was 0.75 tonnes of firewood per tonne of soap. Soap production, therefore, required about 3 tonnes/day.

Other consumers

Choultries: There were 297 choultries in Bangalore and all of them used firewood for cooking the food served at marriages and other functions. The firewood consumption of choultries can be estimated in two ways. Firstly, there are about 120 auspicious days in a year. It can be assumed that on each of these days about 250 meals are cooked with a firewood consumption of about 1 kg/meal. This corresponds to a total firewood consumption of 8,910 tonnes/year. Secondly, a survey of 15 choultries revealed that each choultry used about 30 tonnes/year. On this basis, about 9,000 tonnes/year or 24 tonnes/day of firewood is utilised by the choultries in Bangalore.

Hostels: The hostels in Bangalore can be divided into three types: Type I for college students and working people, Type II in schools and convents; and Type III in government-aided institutions. All the hostels, barring a few exceptions, used firewood as fuel for cooking and water-heating, but the differences in the food prepared in the three types of hostels resulted in differences in the per capita firewood consumption. A survey was, therefore, carried out in 16 hostels. It was found that the per capita firewood consumption was 1.5 kg/day in Type I hostels, 1 kg/day in Type II hostels and 0.75 kg/day in Type III hostels. From these norms, and the number of inmates in these three types, the daily firewood consumption worked out to be about 24 tonnes/day.

Cremation grounds: There were 48 cremation grounds, and it was estimated that about 20 corpses were cremated every day. The firewood requirement per cremation was one bullock-cart load, i.e. about 0.5 tonne. Hence, the firewood consumption by the cremation grounds was about 10 tonnes/day.

Canteens in industries: There were about 35 large canteens in the major industries of the city. Apart from two or three modern establishments, others used firewood in addition to gas and electricity. The number of users of these canteens was about 36,000 persons/day and the per capita firewood consumption was 0.125 kg/day. Hence the total firewood consumption was about 4.5 tonnes/day.

Road asphalting: About 100 drums of tar were heated every day for the asphalting of roads, and each drum required about 40 kg of firewood. This corresponds to a total daily consumption of about 4 tonnes of firewood per day for road asphalting.

It is now possible to assemble the results regarding the consumption of firewood by the different categories of consumers in Bangalore. The pattern of firewood consumption in the city is shown in Figure 4.3.

It can be seen from the figure that, firstly, Bangalore consumes about 1,280 tonnes of firewood every day and, secondly, about 78 per cent of the total consumption is by households. The next major consumer is industry (cottage, medium and large) with a consumption of 9 per cent, with the remaining consumed by restaurants, hostels, etc.

End-use analysis

Firewood is used for cooking, water-heating and for process heat. Water-heating utilised 59 per cent of the total firewood consumed in Bangalore, and is the major end-use,

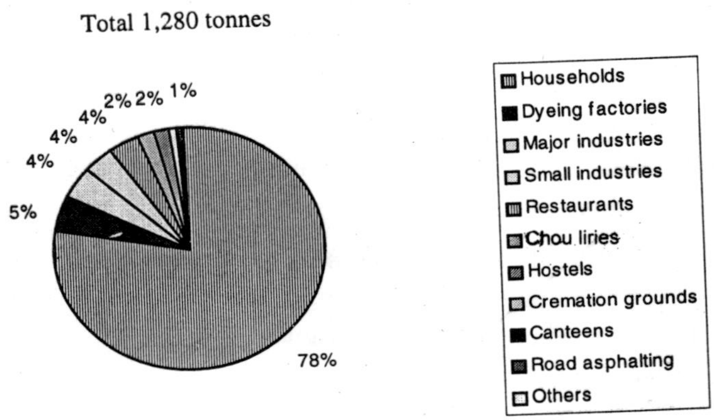

Figure 4.3: Consumption of firewood by various categories.

followed by cooking which utilised 36 per cent. The remaining 3 per cent is used for process heat (Figure 4.4).

Discussion

Implications of Bangalore's firewood supply and consumption

(i) *Impact on forests*: There is a fundamental difference between the usage of firewood in cities and towns and that in villages. Rural households depend overwhelmingly on twigs and branches. A study of villages in Kunigal and Tumkur districts (ASTRA,1982) showed that most of the households depended on twigs, roots and branches obtained predominantly

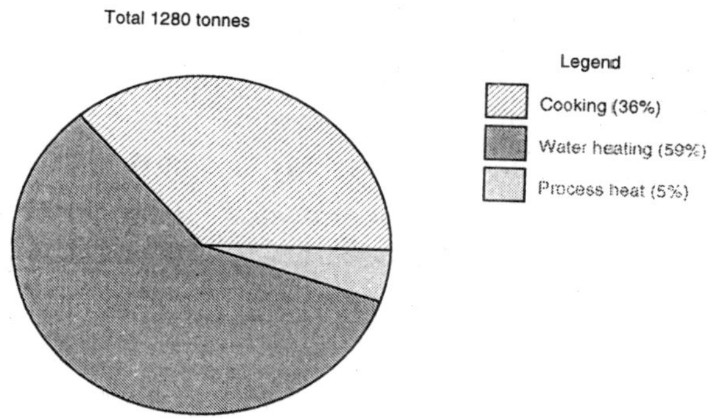

Figure 4.4: End-use analysis of firewood consumption.

by gathering. In contrast, Bangalore, like all cities, uses logs which usually require the felling of trees. The urban preference for logs with a high bulk density rather than twigs with a low density is a consequence of the need to transport the firewood over large distances. Thus, the urban consumption of firewood has a much greater negative environmental impact than the rural dependence on firewood as a domestic fuel.

The procurement of Bangalore's firewood is most probably achieved in a non-renewable manner by the felling of trees. If that is the case, it corresponds to the deforestation of about 3,650 hectares of good forest per year assuming that the growing stock per hectare in such a forest has a high value of 120 tonnes/ha (Anantha Narayana,1988). In other words, about 10 hectares of the best forest would have to be cleared every day to maintain the firewood supply to Bangalore.

(ii) *Impact on transport system*: Transportation of 1,200 tonnes/day of firewood (reported in Section 4.1.2) involved an average of 114 trucks and 10 wagons per day. These numbers must be compared with the records of weigh-bridges (for trucks) and railway goods sheds (for railway wagons) for the months of October-November, 1984 (Anon, 1984) that the total number of trucks and railway wagons bringing freight into Bangalore every day was 630 ± 81 and 125 ± 37.

It follows that about 18 per cent of the truck and 8 per cent of the railway wagon traffic into Bangalore is committed to firewood transport. Thus, in a situation where transport is often a constraint on the economy, a significant fraction of both road and rail transport is tied up with the supply of firewood. There is an "opportunity cost"[1] with this firewood transport. Often, the opportunity cost is measured by the money paid, since this expenditure denies the opportunity to spend the money on something else. In the present case, as already mentioned, transport is often a constraint on the economy and, therefore, instead of transporting firewood, other essentials like foodgrains, medicines, etc, could have been transported. Thus, the cost that society is incurring by not transporting the essential items is considered as the opportunity cost measuring associated with the withholding of this truck and wagon capacity from the economy, but this opportunity cost has not been calculated here.

(iii) *Energy costs of firewood supply:* Since large quantities of firewood are being transported into Bangalore by trucks and railway wagons, it would be interesting to estimate the fuel and energy costs of firewood supply to Bangalore.

(a) Trucks: It has been mentioned that 35 per cent of the firewood came from within a radius of 30-40 km. This supply was by trucks, and corresponds to about 14,700 tonne km/day. Another 50 per cent also came by truck from places on both sides of the Andhra-Kolar district border about 135 km from Bangalore. This component corresponds to about 81,000 tonne km/day. Finally, of the 6-7 per cent firewood coming from a distance of about 300-400 km, only about one-third can be considered to be transported via truck and this corresponds to about 8,400 tonne km/day. All this constituted a total of about 104,100 tonne km/day which, at an average of 22.04 tonne km/kg diesel or 18.80 tonne km/litre of diesel (*Encyclopedia of Energy*,1971), required 4.7 tonnes diesel/day (1,724 tonnes diesel/year).

---

[1] Opportunity cost of goods or services refers to the cost of forgoing/adopting that goods or services and choosing/avoiding another goods or service requiring the same resources (*Encyclopedia of Energy*, 1971)

(b) Railway wagons: On an average, about 56 tonnes/day of firewood was transported via the Southern Railway from Birur junction, 212 km from Bangalore. This corresponds to 11,872 tonnes km/day. In addition, the South Central Railway transports about 60 tonnes of firewood/day from around Goa, which is 680 km from Bangalore. This corresponds to 40,800 tonne km/day, leading to a total of 52,672 tonne km/day for both the railways involved in Bangalore's firewood transport. Using the consumption norm of 154 tonne km/kg diesel, or 131 tonne km/litre diesel (*Encyclopedia of Energy*,1971), the two railways together consume a total of 0.34 tonne diesel/day (125 tonne/year) or 401 litres diesel/day on firewood transport to Bangalore.

Hence, considering both truck and rail transport, the diesel consumption on Bangalore's firewood transport works out to about 5.1 tonnes/day (1,849 tonnes/year or 2.2 million litres/year).

The energy content of diesel is about 37.00 x 106 joules/litre. Hence, the 5.1 tonnes of diesel used every day for firewood transport involve an energy expenditure of 220 x 10^9 joules/day. On the other hand, the 1,200 tonne of firewood/day is burnt with an efficiency of around 10 per cent which means that at the rate of about 16.00 x 10^6 joules/kg of firewood, the energy delivered is about 1.728 x 10^{12} joules/day. Hence, to deliver 100 units of heat energy from the combustion of firewood for the two main end-uses, viz. cooking and water heating, 13 units of energy in the form of diesel fuel have to be expanded on firewood transport.

(c) Foreign exchange costs: The 2.2 million litres of diesel/year utilised for transporting Bangalore's firewood cost about Rs. 32.55 lakhs at an international rate of $14.4 per barrel of crude (Feb. 1990 price) or Rs.1.48/litre of diesel. Since about one-third of India's oil is imported and paid for in foreign exchange, it can be argued that Rs.10.88 lakhs or about $60,000/year is the foreign exchange expenditure on supplying Bangalore with its firewood requirements.[2]

Forest Department Control over Firewood Problem

The use of significant quantities of firewood in Bangalore has been argued above to have four important impacts on the environment and economy: (1) deforestation, (2) reduction of the capacity of the transport system, (3) diesel consumption, and (4) foreign

[2]1 US$=Rs. 18 in 1990.

exchange outflow. All this means that the consumption of firewood is a serious problem which cannot be ignored because it can only worsen unless it is timely tackled.

At the outset, therefore, it is worth considering the possibility of control over the supply and distribution of firewood. Unfortunately, the present study has shown that the government/public sector has only a marginal control over firewood supply. The Forest Department is responsible for Karnataka's forests but supplies less than 5 per cent of Bangalore's firewood, the Railways transport only about 10 per cent, and the cooperative societies handle only 5 per cent of the firewood distribution. In contrast, private contractors supply 95 per cent of the firewood, privately owned trucks carry 87 per cent of the firewood and retail depots handle 85 per cent of the firewood sales. Thus, except for the growth of firewood for which the responsibility rests largely with the government (Forest Department), the remainder of the firewood fuel-cycle involving the extraction, transport and distribution is mainly in private hands and is controlled by the contractor-trucker-retailer nexus. It is also clear that, *in the urban context*, firewood is not non-commercial energy – it is as commercial a form as electricity, oil or coal. In such a situation, it is not easy for the Forest Department to exert any leverage and achieve its objective of rationally managed forests. There appears to be little chance of the Forest Department controlling the situation at the supply-transport-distribution parts of the firewood fuel-cycle.

A demand management approach focussed on end-uses

The chief option, therefore, seems to involve solutions based on the consumption end of the fuel-cycle by highlighting the real origins of the firewood problem.

This task has been simplified by the results on the distribution of firewood consumption over the different categories of consumers from which it is clear that 78 per cent of the total firewood is utilised for domestic consumption, and 79.2 per cent of this firewood is consumed by households where the per capita monthly income is less than Rs.200 per month. The solution to the problem, therefore, lies in providing an inexpensive and efficient cooking fuel and device for cooking and water-heating to the poorer households in cities and towns. Fortunately, the technological component of this solution can be achieved in several ways.

***Efficient firewood stoves for cooking and water-heating*:** The whole question of the efficiencies of firewood stoves used by an overwhelming majority of the country's households used to be largely ignored by scientists and engineers. Over the past decade or two, however, there has been an increasing interest in firewood stoves as a crucial part of the concern over rural energy. It rapidly became clear that efficiencies as high as 30 per cent can be achieved in comparison with the present 10 per cent, i.e. only about one-third of the present input of firewood is required to deliver the same heat energy to the cooking pot. Since about one-third of the firewood used in Bangalore is for cooking, it means that what now requires about 514 tonnes/day can be achieved with as low as 154 tonnes/day.

When the locations at which cooking and water-heating are carried out are close together, it is possible to utilize the waste heat from the cooking for heating water. This cascading[3] approach in which the unutilized energy from one task is used for another task is particularly convenient in firewood stoves which can be designed so that on its way to the chimney the waste heat from the cooking section of the stove can be used to heat water. This would mean that in households which use firewood for both cooking and water heating, efficient stoves can be installed which accomplish both tasks in a single process.

Efficient water-heating: Since cooking cereals requires the boiling of water, efficient water-heating can be achieved with the same design approach as that used for cooking stoves. In other words, water-heating can be achieved with efficiencies as high as 30 per cent compared to the present 10 per cent.

Immediate reduction of firewood consumption: These design improvements can lead to a substantial reduction in firewood consumption which indicate that Bangalore's demand of firewood can be brought down to 513 tonnes/day from the present 1,280 tonnes/day (Table 4.11).

Such a reduction requires a large-scale effort to disseminate improved firewood stoves and water-heaters. One way of achieving this objective is to mass-produce and market, perhaps at subsidised prices, entire stoves or at least those critical parts of stoves which require rigid adherence to dimensions. Flat-plate collectors of solar energy can also play a role in reducing the usage of firewood for water-heating. But, the metallic

---

[3]In this approach, the waste-heat from cooking is used for water-heating in the households which use firewood for cooking and water-heating.

Table 4.11: Firewood consumption after efficiency improvements.

Consumer type	Firewood consumption (tonnes/day)			
	Present	With improved efficiencies	With cascading approach	With solar water-heater
Households	995.0	298.5	192.6	92.4
Dyeing factories	62.0	62.0	62.0	62.0
Bakeries	55.0	55.0	55.0	55.0
Restaurants	48.0	14.4	14.4	14.4
Industries	50.0	50.0	50.0	50.0
Choultries	24.0	7.2	7.2	7.2
Hostels	24.0	7.2	7.2	7.2
Cremation grounds	10.0	10.0	10.0	10.0
Canteens	5.0	1.5	1.5	1.5
Road asphalting	4.0	4.0	4.0	4.0
Soap factories	3.0	3.0	3.0	3.0
Total	1280.0	512.8	406.9	306.7

solar water-heaters are in the same first-cost bracket as electrical water-heaters, which means that they are likely to be accessible mainly to households with monthly incomes of above Rs. 500. If the potential for solar energy for heating is to be taken advantage of, it is essential that the first cost of solar water-heaters is brought within the Rs 2500-5000 range (the affordable level for a common man). Perhaps the only way this can be achieved is by the use of black plastic water-pillows which act as both collectors and water tanks.

Long-term measures: Green belts and sludge/producer gas

If high-efficiency firewood-based cook-stoves and water-heaters and solar water-heaters are introduced, the demand for firewood may be drastically reduced, but not eliminated. In the case of Bangalore, the introduction of these devices has been estimated to lead to a demand of about 307 tonnes/day with sòlar water heaters. This reduced demand must be sought to be met from energy forests managed in the renewable mode so that the extraction rate is matched to the growth rate.

There are several species of fast-growing firewood crops which are excellent candidates for such energy forests. Their yields vary, but a figure of 10 tonnes/ha/year (Paul, 1981) can be assumed for the present discussion. This figure implies that a reduced demand of 307 tonnes/day or 89,425 tonnes/year requires an energy forest of 8,935 ha or 89.4 sq km to supply Bangalore with firewood in a sustained manner.

Obviously, if the aim is to reduce diesel consumption for transporting firewood to Bangalore, it would be rational to locate the energy forest as close to Bangalore as possible. One approach is to grow a forest belt around Bangalore's circumference[4]. Using the proposed metropolitan area of Bangalore of 321 sqkm which, assuming a circular city, corresponds to a circumference of about 63.5 km, the width of the belt turns out to be about 1.0 km. Thus a 1.0-km-wide belt around Bangalore's proposed metropolitan area would provide its requirements of firewood in a sustainable manner.

Such forest belts around Bangalore and other cities would eliminate the four negative impacts of the present pattern of firewood consumption, viz. deforestation, reduction of the capacity of the transport system, diesel consumption, and foreign exchange outflow. If, however, the firewood from these forest belts is used directly in stoves and water-heaters, one major socio-economic problem will persist; different sections of society will continue to use different fuels, the rich will use LPG for cooking and the poor will use firewood, albeit with improved stoves. There is a sound technical reason for the preference of the rich–gas-cooking is far more convenient due to quick lighting of the fuel and the ease of turning the heat output up or down. It may be necessary at this stage to replace the direct supply of firewood from the forest belt with the gasification of wood to generate producer gas which can then be piped to homes. If such a piping system is installed, the producer gas can be supplemented with methane-rich biogas obtained from the treatment of the city's sewage.

Charcoal

The Production

Charcoal is obtained when wood is heated with limited access to air to temperatures of about 400^0 C. Almost all species of wood can be carbonized to give charcoal, but it requires mature dense wood to produce good charcoal. In contrast, rotten and old wood gives a light product, and stems and branches yield soft bulky charcoal (Anon, 1948). Charcoal can be produced in distillation plants where by-products such as wood-gas, methanol, acetic acid and wood tar are recovered. But, it can also be produced in simple

---

[4] The opportunity cost of 63.5 sq.km land on Bangalore's periphery is not considered here.

kilns made of metal, brick-work or even earth-work. The charcoal supplied to Bangalore is produced by traditional methods in either above-ground stack kilns or underground pit kilns. For both these types of kilns, a layer of grass or leafy branches and earth inhibits air infiltration into the kiln. Table 4.12 provides a list of plant species that are being used to produce the charcoal which is supplied to Bangalore.

It appears that the species, *Prosopis juliflora*, is the predominant source accounting for about one-fourth of the total charcoal production.

The Supply

Charcoal is supplied to Bangalore not only from locations within the State but also from areas in the neighboring States of Tamil Nadu and Andhra Pradesh. According to the avaiable information, about half of Bangalore's charcoal comes from Andhra Pradesh; about 70 per cent from distances of over 200 km and the rest from Chickamagalur in Karnataka (250 km) and Dhankinakota in Tamil Nadu (70 km). The suppliers of charcoal fall into two categories: (i) manufacturers of charcoal who supply to the distributors in Bangalore; (ii) contractors who purchase charcoal from the manufacturers and supply it to the distributors.

The data obtained from the suppliers indicate that, during the year 1980, the supply was 53,060 tonnes which decreased to 43,020 by the year 1990 (Figure 4.5). This supply involved a monthly transport of 3980 ± 409 tonnes in 567 trucks. This would correspond to an average daily inflow of 132 tonnes of charcoal in about 19 trucks.

Table 4.12: The species being used to produce charcoal.

Botanical name	Trade name	Local name	Usage (%)
Prosopis juliflora	Vilayate babool	Bellarijali	25
Aegle marmelos	bel	Bilipatre	
Anogeissus latifolia	axlewood	Dinduga	
Azadirachata indica	neem	Bevu	
Acacia nilotica	babul	Jali	30
Acacia leucophloea	reunjha	Bilijali	
Haldina cordifolia	haldu	Yettiga	
Butea monosterma	palas	Muttuga	
Mangifera indica	mango	Mavu	30
Tamirindus indicus	tamarind	Hunase	
Trewia mudiflora	gutel	Katkambala	
Terminalia alata	laurel	Mathi	15

The Transportation

Trucks were the only mode of transportation used for carrying charcoal to Bangalore. Charcoal is filled in bags and loaded onto trucks for transport. The number of bags transported per truck was 215 ± 2.

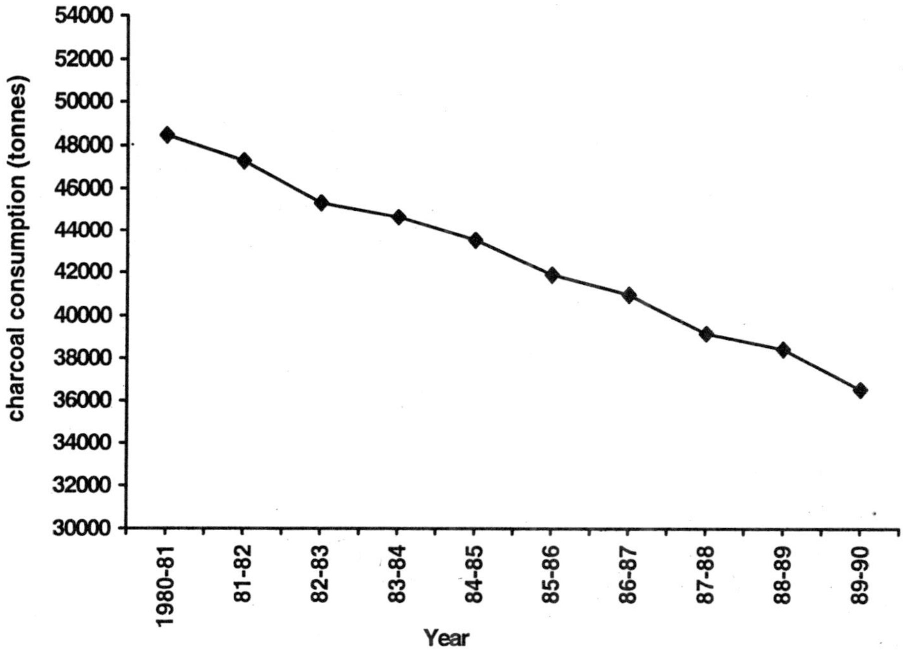

Figure 4.5: Supply of charcoal to Bangalore.

The Distribution

The distribution of charcoal is through 12 commission agents in Bangalore who receive charcoal either from manufacturers or from private contractors and distribute it to the 376 charcoal depots. They also make direct supplies to industries and to some hotels. Other consumers purchase directly from the retail depots (Figure 4.6).

Nearly 75 per cent of the charcoal supplied to Bangalore was through contractors while the direct supplies from manufacturers accounted for 25 per cent. The commission agents, who receive all this charcoal, in turn channel 80 per cent of the total inflow to retail depots They supply the remaining 20 per cent directly to bulk consumers.

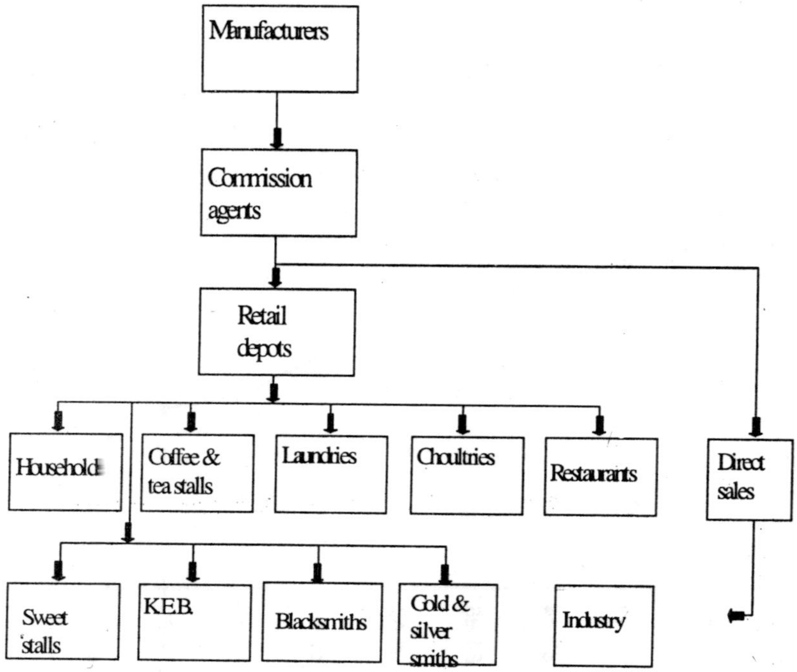

Figure 4.6: Channels of supply of charcoal to Bangalore.

The Consumption

The consumers of charcoal in Bangalore can be classified into the following categories:(1) residential, (2) commercial, (3) industrial and, (4) other consumers.

Residential consumption

The number of households using charcoal in different income groups is given in Table 4.13 from which it can be seen that 92 households, i.e. 9.2 per cent of the sample, use charcoal either wholly, or partially along with another fuel. Of these 92 households, 27 households depend wholly on charcoal and use it for both cooking and water-heating, in contrast to 39 households who depend on charcoal for cooking only, and 26 households for water-heating only.

The table also shows that there is an income dependence in charcoal consumption. The percentage of households (in a particular income range) that use charcoal either wholly or partially varies with income range. In particular, the percentage of households using charcoal seems to rise slightly with income and then declines. In other words, charcoal consumption seems to be concentrated among the middle-income groups rather than among the poorest or richest sections.

The dependence of charcoal consumption on income can be revealed through the definition of a charcoal dependence index (TCDI) as a ratio of the percentage of households dependent on charcoal to the percentage of households in that range (Table 4.14).

The TCDIs for various income groups are presented in Table 4.14 from which it can be seen that households with a per capita monthly income of less than Rs 200 bear a greater burden of charcoal consumption than their percentage in the sample. Incidentally, these households which constitute 42.8 per cent of the sample of 1,000 households account for 81.4 per cent of the households which are totally dependent on charcoal. The TCDI decreases with an increase in income according to the regression equation

$$\text{TCDI} = 2.2151 - \underset{(0.0023)}{0.0054}\, X\,(\text{mid-point of PCMI range}) \qquad (4.4)$$

$R^2 = 0.95.$

(*Note:* Number in parenthesis represents standard error).

Table 4.13: Pattern of charcoal consumption in the sample.

PCMI (Rs)		No of households in the range	Households using charcoal			
Group	Range		Total	C & WH	C only	WH only
I	<100	148 (14.8)	14 (15.2)	8 (29.6)	4 (10.3)	2 (7.7)
II	100-149	168 (16.8)	20 (21.7)	10 (37.0)	6 (15.4)	4 (15.4)
III	150-199	112 (11.2)	17 (18.5)	4 (14.8)	8 (20.5)	5 (19.2)
IV	200-249	121 (12.1)	17 (18.5)	2 (7.4)	8 (20.5)	7 (26.9)
V	250-299	120 (12.0)	14 (15.2)	2 (7.4)	7 (17.9)	5 (19.2)
VI	300-349	138 (13.8)	7 (7.6)	1 (3.7)	4 (10.3)	2 (7.7)
VII	>350	193 (19.3)	3 (3.3)	-	2 (5.1)	1 (3.9)
Total		1000 (100.0)	92 (100.0)	27 (100.0)	39 (100.0)	26 (100.0)

Note: Figures in parentheses represent percentages.

Table 4.14: Total charcoal dependence index.

PCMI (Rs)		Mid-point of income range	% of households in the sample	% of households dependent on charcoal for C&WH	TCDI (col 5/col 4)
Group (1)	Range (2)	(3)	(4)	(5)	(6) =(5)/(4)
I	<100	72	15.0	29.6	2.00
II	100-149	121	16.8	37.0	2.18
III	150-199	172	11.2	14.8	1.35
IV	200-249	222	12.1	7.4	0.61
V	250-299	272	12.0	7.4	0.62
VI	300-349	325	13.8	3.7	0.27
VII	>350	382	19.3	-	-

Another way of showing the income dependence of charcoal consumption is by drawing the Lorenz curve (cumulative percentage of charcoal-using households *vs* cumulative percentage of households) and observing the departure of the curve from the 45^0 slope straight line representing equal distribution of the charcoal burden. The Lorenz curve (Figure 4.7) shows greater concentration of totally charcoal-dependent households amongst the low- and middle-income groups.

The households which consume charcoal for cooking fall into two categories: those which consume charcoal for both cooking and water-heating and those which consume it for only cooking (heating water either with firewood, electricity or dung), using kerosene (KRC) and dung (DGC) as auxiliary fuels. Charcoal consumption (CHC) (kg per month) was regressed on family income (FI) (Rs per month) and family size (FS) (no.) for the first category and for the second one, charcoal consumption was regressed on family income, family size, and also on the consumption of kerosene and LPG. The resulting equations are:

For cooking and water-heating,

$$CHC = 2.84 + \underset{(0.94)}{9.83\, FS} + \underset{(0.005)}{0.0178\, FI} \tag{4.5}$$

$$R^2 = 0.85$$

For cooking,

$$CHC = 5.1 + \underset{(0.82)}{4.80\, FS} + \underset{(0.003)}{0.02\, FI} - \underset{(0.14)}{0.12\, KRC} - \underset{(0.62)}{0.21\, DGC} \tag{4.6}$$

$$R^2 = 0.85$$

(*Note:* Numbers in parentheses represent standard errors).

In the case of cooking and water-heating, the regression coefficient of family income contributes less than 2 per cent of the increase in charcoal consumption. In the

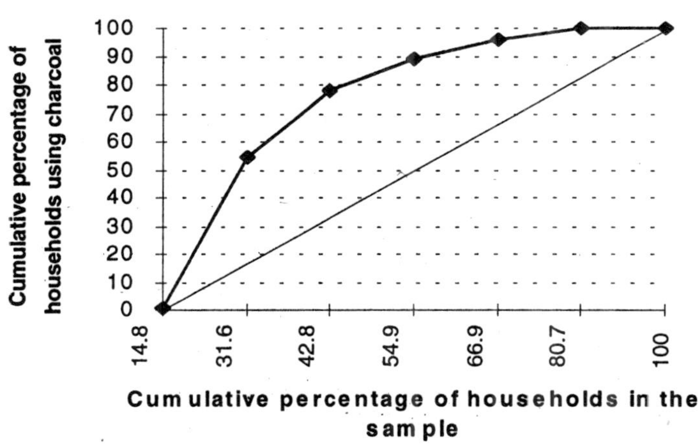

Figure 4.7: Lorenz Curve for charcoal usage.

case of charcoal consumption for cooking only, again the coefficient of family income is very low. Also, charcoal consumption decreases with an increase of kerosene and dung consumption.

From Table 4.15, the following conclusion can be inferred: within the category of households that use charcoal, family income does not play any significant role in the consumption of charcoal. Though this conclusion appears perplexing, it is reasonable because only households in a particular income range consume charcoal, and income stratification is unlikely within this range. Hence, within a particular income range, charcoal consumption depends much more on family size than on family income. The *t* values indicate that the regression coefficients are significantly different from zero and the values of R^2 (for both the equations) indicate that the variables included in the model explain 85 per cent of the variation in charcoal consumption.

Table 4.15: Regression results.

Variable	Mean	Std.dvn	Variable	Coefficient	Std.dvn	't' value	R^2 (individual)	R^2 (step-wise)
C&WH								
Family size (no.)	5	1	FS	9.830	0.940	10.46	0.72	0.72
Family income (Rs/m)	199	79	FI	0.036	0.005	7.20	0.18	0.85 (FS+FI)
Cooking only								
Family size (no)	5	0	FS	4.800	0.820	5.85	0.69	0.69
Family income (Rs/m)	202	91	FI	0.020	0.003	6.67	0.11	0.77 (FS+FI)
Kerosene consumption (l/month)	3	2	KR	-0.120	0.140	-0.86	0.10	0.83 (FS+FI+KR)
Dung consmn (kg/month)	10	9	DG	-0.210	0.620	-0.34	0.03	0.85 (FS+FI+KR+DG)

Thus far, the description has been restricted to the number (and percentage) of charcoal-using households. If the quantities of charcoal are considered (Table 4.16) it is seen that the average per capita charcoal consumption (kg per year per capita) for cooking only and for water-heating only, viz. 111.3 and 75.3 kg per year per capita respectively, is greater (13 per cent) than the corresponding figure, i.e. 160.2 kg per year per capita, for cooking plus water-heating. This suggests that if cooking and water-heating are done separately, the energy efficiency is less compared to the situation when they are done simultaneously. Secondly, there appears to be a definite increase in the per capita charcoal consumption with per capita monthly income. The per capita consumption of Group VII is 24 per cent higher than in Group I. In fact, the per capita charcoal consumption is linearly correlated with the mid-point of the per capita monthly income range, the coefficient of correlation being above 0.97.

If the per capita expenditure on charcoal is considered, then the percentage of the per capita monthly income spent on charcoal consumption decreases with an increase in the per capita monthly income, and the poor have to spend a greater percentage of their earnings on charcoal. This trend is found irrespective of whether the charcoal consumption is for cooking and water-heating, or for cooking only or for water-heating only. For example, in the case of charcoal for cooking and water-heating, with a mean per capita monthly income group of Rs 72, the households in Group I would have to spend as much as 21.4 per cent of their monthly income on charcoal compared to only 5.7 per cent for group VI (Table 4.17).

Table 4.16: Per capita consumption of charcoal for cooking and water-heating.

PCMI (Rs.)		Mean household size	Consumption (Kg/year)					
			Cooking & Water-heating		Cooking only		Water-heating only	
Group	Range		Per household	Per capita	Per household	Per capita	Per household	Per capita
I	<100	5	711.8	136.1	376.6	72.0	276.1	52.8
II	100-149	5	755.0	156.0	464.6	96.0	273.0	56.4
III	150-199	5	725.2	160.8	487.1	108.0	297.7	66.0
IV	200-249	4	711.2	162.0	511.0	116.4	326.6	74.4
V	250-299	4	715.7	168.0	541.9	127.2	357.8	84.0
VI	300-349	4	725.9	178.8	531.0	130.8	375.1	92.4
VII	>350	4			491.8	128.4	386.1	100.8
Average		4.4	724.1	160.2	486.3	111.3	327.5	75.3
Std.deviation		0.4	15.0	13.0	51.1	19.8	43.3	16.8
Var.coefficient (%)		9.1	2.0	8.0	10.5	17.8	13.2	22.3

The extrapolation of the results obtained from the 1,000-household sample of the city of Bangalore has been achieved in the following way. Firstly, it has been assumed that the 1000-household sample represents the whole city with regard to the percentages of households in any income range using: (a) charcoal for both cooking and water heating, (b) charcoal for cooking only and (c) charcoal for water heating only. Secondly, the per capita charcoal consumption norms observed in the sample have been considered to be valid for the whole of Bangalore. Thirdly, since the income distribution for the 1,000-household sample may not be a true representative of Bangalore, the study of Prakasa Rao and Tewari on the economic structure of Bangalore has been used to get the

1973-74 picture of the percentages of households in various income groups of per capita monthly income. These percentages have been updated (to 1983) to allow for increases of income since 1974. As emphasised in the firewood section, this procedure assumes that the income distribution in Bangalore has not changed over the past decade. Even if this were to be different, in view of the lack of data, the distributional changes in income could be presumed to be too small to warrant serious consideration at this stage.

Table 4.17: Per capita expenditure on charcoal consumption for cooking and water-heating.

PCMI (Rs)		Mid-point of income range	Cooking and Water-heating		Cooking only		Water-heating only	
Group	Range		PCME (Rs)	As % of income	PCME (Rs)	As % of income	PCME (Rs)	As % of income
I	<100	72	15.4	21.4	7.5	10.40	5.50	7.6
II	100-149	121	16.25	13.4	10.0	8.26	5.87	4.8
III	150-199	172	16.75	9.74	11.25	6.54	6.87	4.0
IV	200-249	222	16.90	7.60	12.12	5.46	7.75	3.5
V	250-299	272	17.50	6.43	13.25	4.87	8.75	3.2
VI	300-349	325	18.62	5.73	13.62	4.19	9.60	2.9
VII	>350	382	-	-	13.34	3.50	10.5	2.7

On this basis, Bangalore's charcoal consumption can be estimated to be 15,228 tonnes per year for households which use only charcoal for both cooking and water-heating, 12,837 tonnes per year for households which use charcoal for cooking only and 5,610 tonnes per year for households which utilize charcoal for water-heating only. This adds up to a total of 33,675 tonnes per year which would correspond to an average daily consumption of about 92.5 tonnes of charcoal for the domestic sector (Table 4.18). It is seen that households with a per capita income of less than Rs. 250 per month, which constitute about 78 per cent of Bangalore's households, are estimated to consume about 86 per cent of its charcoal consumption. Hence the charcoal consumption appears to be shared in a reasonably egalitarian manner, but for the fact that groups with a per capita monthly income greater than about Rs. 300 use a smaller share of charcoal than would be indicated by their share of the population.

Table 4.18: Contribution of various income groups in charcoal consumption.

PCMI (Rs)		HH in the Range		Usage of HH		Consumption (tonnes/year)	
Group	Range	Per cent	Number	Per cent	Number	Total	Per cent
I	<100	31.9	189680	9.45	17925	27.40	29.6
II	100-149	12.7	75515	11.90	8986	14.20	15.4
III	150-199	15.6	92759	15.17	14072	19.00	20.5
IV	200-249	17.6	104651	14.04	14693	18.90	20.4
V	250-299	8.6	51136	11.65	5957	8.30	9.0
VI	300-349	10.0	59461	5.07	3015	4.30	4.6
VII	>350	3.6	21406	1.55	332	0.40	0.4
	Total	100.0	59608		64979	92.50	100.0

Commercial consumption

Restaurants: At the time of the survey, there were 1,082 vegetarian and 787 non-vegetarian restaurants in Bangalore. A sample survey conducted on 5 per cent of the restaurants showed that 60 per cent of the non-vegetarian restaurants used charcoal at an average consumption rate of 30 kg per day. In the case of vegetarian restaurants only 25 per cent of them used charcoal with an average consumption of 20 kg per day. Hence, the daily consumption of charcoal by the hotels in Bangalore can be taken as 19.6 tonnes per day.

Coffee and tea stalls: The number of coffee and tea stalls in Bangalore were 254 and 495, respectively. A sample survey covering 5 per cent of the total number of coffee and tea stalls revealed that 25 per cent of the coffee stalls and 75 per cent of the tea stalls used charcoal. The quantity of charcoal consumed per day was about 10 kg for a coffee stall and 15 kg for a tea stall. Thus, the total consumption of charcoal by the coffee and tea stalls in Bangalore works out to be 6.2 tonnes per day.

Sweetmeat stalls: A sample survey thus covering 5 per cent of the total number of 346 sweetmeat stalls revealed that 60 per cent of them used charcoal with a daily consumption of about 15 kg per stall. The daily consumption by the sweetmeat stalls of Bangalore thus works out to 3.15 tonnes per day.

Laundries: There were about 615 laundries in Bangalore and all of them used charcoal-heated irons. On an average, each laundry consumed about 5 kg per day, and thus the total quantity of charcoal consumed by laundries comes out to about 3.1 tonnes per day.

Industrial consumption

Agarbathi (incense stick) industries: Charcoal powder is used for the preparation of incense sticks in agarbathi manufacturing units of which about 500 are in Bangalore. On an average, each unit consumes about 10 kg of charcoal per day, which means that the consumption of charcoal by the Bangalore's agarbathi units in Bangalore is about 5 tonnes per day.

Other types of industries: According to the Chief Inspectorate of Boilers, two major industries (Bangalore Industries Gases and Chemicals Corporation and Grindwell Norton Ltd.) consume about 2 tonnes of charcoal per day. There are also other small industries that use charcoal for their boilers; their total consumption being about 3 tonnes per day.

Thus the total consumption of charcoal by the industries in Bangalore can be taken as about 10 tonnes per day.

Other types of consumers

Karnataka Electricity Board: While erecting poles for transmission/distribution of electricity, the Karnataka Electricity Board digs pits (to receive the poles) and then fills the pits with charcoal (on an average of 15 kg per pit) for earthing the poles. The number of poles erected and, therefore, the earthing done by the Electricity Board in Bangalore was 3,200 per month. Their consumption of charcoal can, therefore, be taken as 1.6 tonnes per day, but this is a non-energy end-use.

Choultries: There were 297 choultries in the city where marriages and other functions are held on about 120 *auspicious* days in a year. During these functions, charcoal is used for making coffee and tea. The total quantity of charcoal consumed by the choultries was estimated to be about 5 kg per choultry per function day, which works to about 181 tonnes per year, i.e. an average of 0.5 tonnes per day.

Blacksmiths: There were about 97 blacksmiths in Bangalore who used charcoal for their furnaces. Each smith used about 2 kg of charcoal in addition to rice husk and dung cakes. This means that the quantity of charcoal consumed by a blacksmith comes to be about 0.2 tonnes per day.

Goldsmiths: About half of the 214 goldsmiths in Bangalore used charcoal to melt gold and silver while the other half used rice husk. Their daily consumption was about 1 kg per smith, resulting in a total charcoal consumption of about 0.2 tonnes per day.

Total charcoal consumption

If the consumption of different categories of consumers of charcoal in Bangalore is added, a figure of about 137 tonnes per day is obtained (Figure 4.8).

Further, in the total consumption, households account for about 70 per cent and the domestic, hotel and industry sectors together constitute 90 per cent of the charcoal used in Bangalore with all the other categories constituting a mere 10 per cent.

End-uses of charcoal consumption

The consumption by various categories of charcoal consumers can be disaggregated into the following types of end-uses: (i) cooking, (ii) water-heating, (iii) process heating and (iv) miscellaneous. The results show that 58.2 per cent of the total charcoal goes for cooking, 30.8 per cent for water-heating and the remaining 11 per cent for process heat and for miscellaneous non-energy end-uses such as electrical earthing.

Discussion

Impact on forests

The efficiency of conversion of oven-dry wood into charcoal is about 30 per cent in the type of kilns producing the charcoal used in Bangalore. The average daily charcoal requirement of the city was about 137 tonnes which implies that about 473 tonnes of wood have to be carbonised every day to meet Bangalore's requirements. Assuming a liberal estimate of about 120 tonnes of wood per hectare of cleared forest, it seems that about 4 hectares of forest would have to be cleared each day to keep Bangalore supplied with its charcoal demand if the wood is being obtained non-renewably by clearing forests.

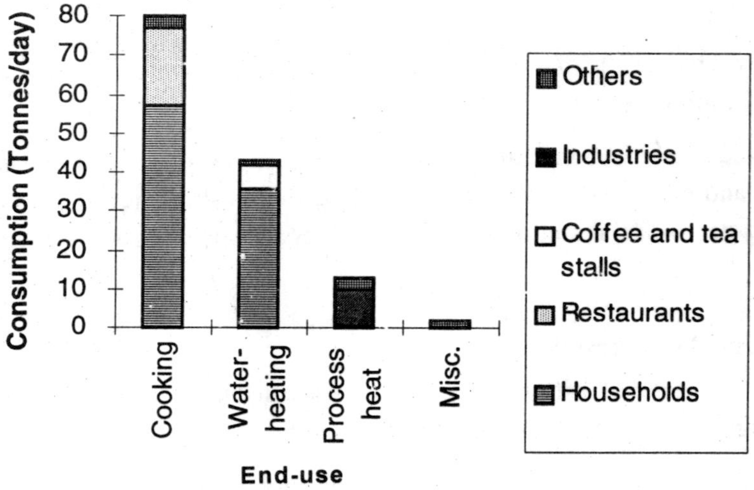

Figure 4.8: Pattern of charcoal consumption.

Impact on transport system

Charcoal is transported entirely by trucks On an average, about 19 trucks per day, i.e. approximately 3 per cent of the 630 trucks bringing freight into Bangalore are involved in this transport.

Energy cost of charcoal supply

It is of interest to estimate the energy expenditures associated with the supply of charcoal to Bangalore. Of the total charcoal supplied to Bangalore, 36.5 per cent comes from a distance of 220 km which corresponds to 10,600 tonne km per day. Another 32.9 per cent (which corresponds to 10,857 tonne km per day) comes from the forests of Chickamagalur which are at a distance of 250 km from Bangalore. And 16 per cent of charcoal comes from nearby forests which are at a distance of 70 km. This corresponds to about 1,478 tonne km per day. Finally, 14.6 per cent of charcoal comes from a distance of 180 km, and this corresponds to 3,469 tonne km per day. All this constitutes about 26,404 tonne km per day. Using the consumption norm of a truck as 18.80 tonne km per litre of diesel, 1.404 kl of diesel have to be used every day to bring in Bangalore's charcoal. This corresponds to an annual requirement of about 512 kl of diesel.

Further, the energy content of the average daily consumption of 1.404 kl of diesel is 51.95 x 10^9 joules per day on the basis of 37.0 x 10^6 joules per litre. This expenditure of energy leads to the supply of 132 tonnes of charcoal which have an energy content of 4.37 x 10^{12} joules, taking the calorific value of charcoal as 33.10 x 10^6 joules per kg. But not all this energy which is released upon the combustion of charcoal is utilised for the desired end-use, e.g. cooking. Assuming 15 per cent thermal efficiency of the stoves and other end-use devices, the energy delivered by charcoal combustion is 655.50 x 10^9 joules per day in comparison with 51.95 x 10^9 joules per day of diesel energy used to supply the charcoal. In other words, to deliver 100 units of energy through the combustion of charcoal about 8 units of energy are being expended in the form of diesel.

Foreign exchange costs

The annual consumption of 0.512 million litres of diesel used for transporting charcoal corresponds (at a rate of Rs 1.48 per litre) of diesel to Rs 758,000, one-third of which may be considered to be in the form of foreign exchange for imports. That is, the annual foreign exchange expenditure on diesel for charcoal transport may be taken as about $54,000.

Coal

The supply and distribution

Bangalore's requirement of coal is met mainly from the South-Eastern Coal Fields (Orissa and Madhya Pradesh), Central Coal Fields (Maharashtra) and Singareni Collieries (Andhra Pradesh) which are at distances of 1,000 to 1,500 km from Bangalore. The mining and supply of coal is controlled by the Government of India except in the case of Singareni Collieries, which is run by the Government of Andhra Pradesh. Coal is transported to Bangalore by rail. The supply of coal has decreased in recent years from 184,270 tonnes in 1980-81 to 73,240 tonnes in 1989-90, a decrease of 60 per cent (Figure 4.9).

Coal India Ltd. distributes coal to the railways, major industries and Karnataka Small Scale Industries Development Corporation Ltd (KSSIDC). KSSIDC receives coal from the collieries and allocates it to its member industries according to their requirement. When a consumer's requirement is large, he gets it directly from the collieries.

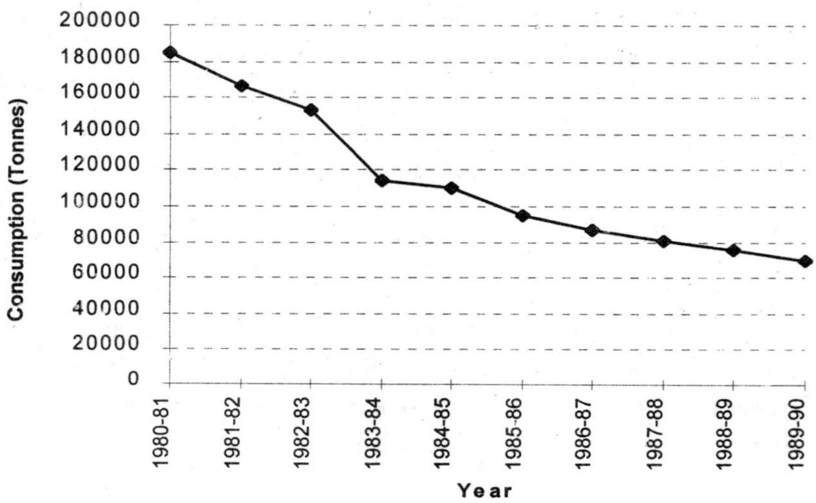

Figure 4.9: Coal consumption in Bangalore (1980-90).

Consumption

The consumers of coal can be broadly classified into two types: (i) railways and (ii) industries. The industries sector consists of the following categories: (a) textile mills, (b) soaps and detergents, (c) chemicals, (d) minerals, (e) iron and steel, (f) brick kilns, (g) fibre industries, (h) engineering industries, (i) public utilities (dairy products, etc.), and (j) food products.

The consumption of coal (CC) has decreased in recent years. The following curve fit measures the trend between 1980 and 1990.

$$CC = 205842 - \underset{(2578)}{2040}\, Y \; (Y = \text{year}) \tag{4.7}$$

$R^2 = -0.95$

The equation shows that the consumption is decreasing linearly over time.

Railways were the major consumers of coal with a daily consumption of around 110 tonnes (1983-84). Their consumption fell from 278 to 47 tonnes between 1980 and 1990. This decrease was mainly due to the dieselisation of railway engines. At present, all the passenger trains in the Bangalore division are dieselised and coal is used only for some goods engines.

In the industrial sector, the textile mills are the biggest consumers of coal followed by Karnataka Soaps and Detergents Ltd (KSDL). There were four major and thirty small textile mills that consumed coal for their boilers. The consumption of coal by these industries has decreased marginally in recent times. On an average, they use about 150 tonnes of coal per day.

Many small-scale industries of different categories in Bangalore consume coal for process heat. During 1984, there were 56 coal-using industries, each with a consumption of around 1 tonne per day. The number of industries in each category and their average per capita consumption are shown in Figure 4.10.

Consumer-wise break-up

Railways with a 44 per cent share of total consumption are the biggest consumers of coal followed by the major industries with a 37 per cent share. The balance is consumed by other small-scale industries.

End-use analysis

Coal is used mainly to produce steam which in turn is used for automotive power and process heat. In the total consumption, the former's share is 44 per cent (by railways) and of the latter 46 per cent.

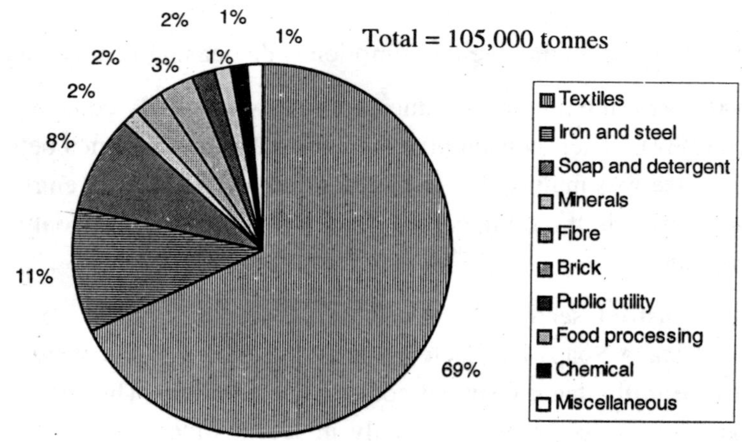

Figure 4.10: Pattern of coal consumption by industries in Bangalore.

Discussion

Coal is by far the most important fossil fuel in India used for power generation and industrial purposes. It is a non-renewable energy source and, therefore, it should be used efficiently.

Bangalore's coal supply is from the Eastern and Central coal fields which are at distances of 1,000 to 1,500 km from Bangalore. Transportation of large amounts of coal is difficult. Also, there are other costs associated with rail delivery such as an adequate coal storage facility during the monsoon period, requirement of wagons for regular coal supply, etc. A train with 40 to 50 wagons of coal satisfies the needs of the city only for 3 days. This is a 1,500 km trip. At an average speed of 50 km/hour (excluding the halts in between), it takes 8 days to receive the coal (*Indian Railways Yearbook*,1981). Thus, 2 to 3 trains should be underway continuously just to keep the city supplied with coal. On an average per day, about 250 tonnes of coal are transported to Bangalore from a distance of 1,500 km., which corresponds to 375,000 tonne-km per day. Another 50 tonnes come from 1,000 km which corresponds to 50,000 tonne-km per day. Thus the total becomes 425,000 tonne-km per day. The consumption norm for a goods train is 131 tonne-km per litre of diesel. This accounts for 3,245 litres of diesel per day. Also, the energy content of the diesel is 37.00 x 10^6 joules. Hence, the 3,245 litres of diesel used every day to transport coal involves an energy content of 120.07 x 10^6 joules per day. On the other hand, 300 tonnes of coal/day is burnt with an efficiency of 35 per cent (*Indian Railways Yearbook*, 1981). This shows that at the rate of 29.30 x 10^6 joules/kg of coal, the energy delivered is 3076.5 x 10^6 joules per day. Hence, to deliver 100 units of heat energy from the combustion of coal, about 4 units of energy in the form of diesel have to be spent on transport. There are also serious impacts of the combustion of coal, such as air pollution through the emissions of carbon dioxide, carbon monoxide, hydrocarbons, etc., which cause damage to the environment. Health is also adversely affected. In addition, Indian coal has a high ash content (30-40 per cent), the disposal of which is problematic (*Indian Railways Yearbook*, 1981). One way to reduce emissions is to employ methods of removing the sulphur during pre-combustion rather than from the effluent. This pre-combustion is cheaper and more effective because the sulphur pollutant is attacked in its more concentrated solid state rather than in a gaseous state. Other sulphur reduction processes such as coal gasification, liquefaction, etc., are also underway, but the cost-benefit analysis of these has to be worked out (Harker *et al.*,1981).

In short, an extensive use of coal is undesirable because of transportation problems and environmental effects which suggest that coal conservation will assume great important in the coming years.

5

PETROLEUM PRODUCTS

The main concern in all scientific work must be the human being himself. This, one should never forget among all those diagrams and equations.

– Albert Einstein

Petroleum products are the dominant source of energy for most of the world. They are the basic source for automobiles and the petrochemical industry. The intensive use of various forms of petroleum has transformed the life of the modern man. The utilisation of petroleum products greatly influences the economic and social status of people. In one way or the other, the consumption of oil has revolutionised our lives in the 20th century.

Increasing use of petroleum products is almost synonymous with the energy crisis. Even though the utilisation of petroleum products has resulted in higher industrial productivity and reduction in our travel time, the demand for energy has increased due to the spatial expansion of society. The conquest of distance by the destruction of time is fuelled by a large increase in energy consumption. There is no doubt that the ever-increasing consumption of oil can be considered as the root cause of our energy crisis. The current debates about climatic change and ozone layer depletion are directly related to the consumption patterns of petroleum products. The automobile is the single largest cause of air-pollution which is playing havoc with the health of humans and all other animals, plants and structures ordinary and of great historical importance. Even with all these negative impacts, petroleum has become indispensable in the life of man. Therefore, understanding these facts is essential from two stand-points. First of all, it should provide an insight into the impact of the increasing consumption of petroleum products and its environmental consequences. Secondly, it is important to study the technological alternatives to increase the efficiency of oil utilisation and substitute other forms of fuel. From this perspective, the consumption of various forms of petroleum products, and their impacts on energy utilisation and environment and the emerging situations need a closer look. The main petroleum products considered in this chapter are: (i) kerosene, (ii) liquefied petroleum gas (LPG), (iii) motor spirit (MS), (iv) high-speed diesel (HSD), (v) light diesel oil (LDO), and (vi) furnace oil (FO).

Supply

Bangalore gets its petroleum products from the Madras refinery which in turn gets its supply of crude oil either from indigenous sources or from oil-exporting countries. The crude is refined and the finished products are supplied to different parts of Karnataka including Bangalore. The finished products are transported to Bangalore through railway wagons and tankers and stored in the storage yards of the respective oil companies.

More than 30 per cent of India's requirements of oil are imported. Iran, Iraq, Saudi Arabia, Kuwait and Russia are the major suppliers of crude oil to India (Venkata Raman, 1990), but the quantity of oil imported from each country varies from year to year depending on the international market price and the indigenous demand.

Transportation

Of the total freight of petroleum products 90 per cent is by rail. Tankers are employed only when railway wagons are not available. This is because the railways are more economical for long-distance transport. On an average, eight to nine wagons per day transport kerosene while the number of tankers is four to five per day. The capacity of a wagon is 30 kilolitres (kl) whereas for a truck it varies from 8 to 12 kl. Usually, 9-kl-tank trucks are used for transporting petroleum products.

In the case of LPG, Bangalore gets its supply from the Mumbai and Koyali refineries which are at a distance of about 1,000 km. LPG is transported through tank trucks often referred to as "bullets" (capacity between 6 and 7 tonnes).

Marketing

Marketing channels

The Government of India allocates petroleum products to Karnataka State every month and the Oil-Coordination Committee re-allocates them in favour of the four oil companies: (i) Indian Oil Corporation Ltd (IOC), (ii) Hindustan Petroleum Corporation Ltd (HPC), (iii) Bharat Petroleum Corporation Ltd (BPC) and (iv) Indo-Burmah Petroleum Corporation Ltd (IBP). The first three are in the public sector and the last is in the joint sector. In Bangalore, IOC supplies 73 per cent of the petroleum products consumed while the share of HPC is 24 per cent and that of BPC and IBP is 3 per cent (1988 figures).

Distribution

Petroleum products are being distributed by the four oil companies which allocate them to 28 wholesale dealers, IOC having 20, HPC 6, and BPC and IBP one each. The wholesale dealers are appointed by the oil companies. After the dealership is granted, they obtain licences to sell petroleum products to the consumers. When the consumption is large, the customers get their requirements directly from the oil companies.

In the case of kerosene licensing, the situation is slightly different. The Joint Director of Food and Civil Supplies (Informal Rationing) is the licensing authority. The Directorate also issues licences to industries and other consumers like restaurants, hotels, etc, for purposes other than illumination. Customers consuming less than 8 kl per month obtain the fuel from wholesale dealers, while the large consumers (consuming more than 8 kl) obtain it directly from the oil companies.

The dealers distribute kerosene to the retailers, registered cooperative societies and to hawkers and vendors who in turn sell it to the domestic consumer (Figure 5.1). While the retailers and registered societies sell kerosene from designated places, the hawkers go from door to door taking kerosene in a barrel on a hand-pulled or a bullock-drawn cart. The hand-pulled cart can carry only one barrel (200 litres) but the bullock-drawn carts carry three barrels. In Bangalore, the number of IOC dealers for LPG was 14 followed by HPC's 12 and BPC's 2. IOC leads in terms of quantity with 65 per cent of the total LPG supply. The share of HPC was 30 per cent and of BPC 5 per cent (Venkat Raman,1990).

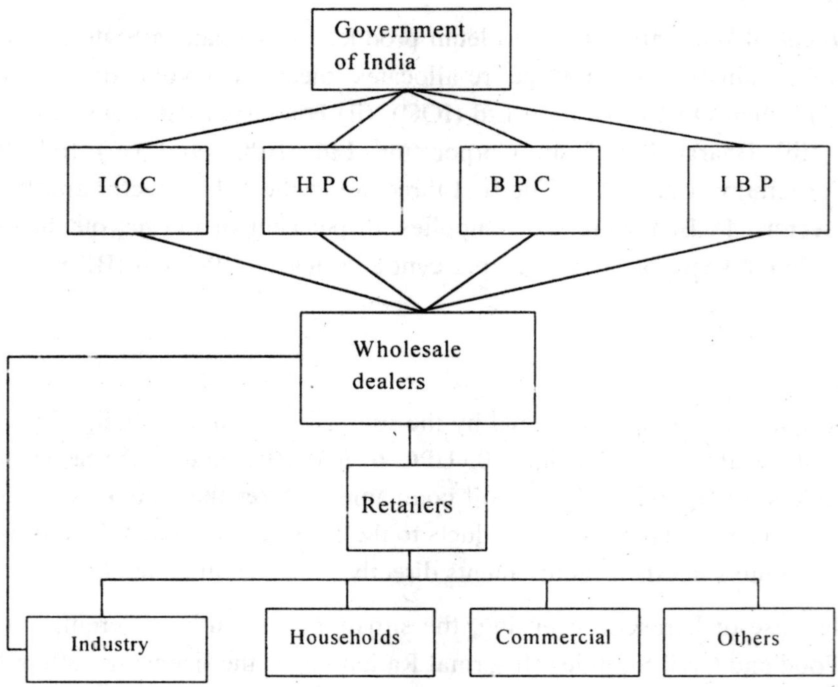

Figure 5.1: Channels of distribution of petroleum products in Bangalore.

The dealers distribute 90 per cent of the total petroleum products while the remaining 10 per cent comes through direct sales to industries. Of the three different channels of distribution, hawkers handle the largest quantity (in case of kerosene). Households prefer the vendors since they sell the fuel near the homes while the retailers and registered societies sell kerosene from fixed locations. Hence, the share of the hawkers in the total distribution was as high as 60 per cent while that of the retailers and registered societies 25 and 5 per cent, respectively.

Kerosene

The consumers of kerosene can be classified into four categories: (i) residential, (ii) commercial, (iii) industrial and (iv) other types of consumers.

The quantity of kerosene consumed in Bangalore is steadily increasing at a rate of 5 per cent per annum. The consumption during 1980-81 was 84,747 kl which increased to 184,343 kl during 1989-90. The supply of kerosene to the domestic customers for a period of eight years is shown in Figure 5.2. The supply was 225 kl per day during 1980-81 which increased to 390 kl during 1989-90, an increase of 66.7 per cent.

The kerosene consumption (KC) is increasing linearly as shown by the following curve fit.

$$KC = 75547 + \underset{(102.5)}{9596}\, Y \ (Y = \text{Year}) \qquad (5.1)$$

$R^2 = 0.97$.

(The number in parentheses represents standard error.)

Household consumption

Based on a household survey the consumers of kerosene can be classified into four different groups. Group I: It consists of families which depend entirely on kerosene for both cooking and water-heating; Group II: For it kerosene is the sole cooking fuel;

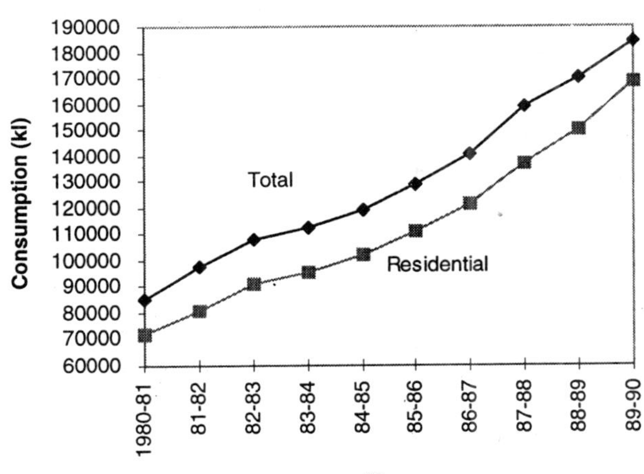

Figure 5.2: Pattern of kerosene consumption (1980-90).

Group III: For cooking it uses kerosene along with either LPG or electricity; and Group IV: It uses kerosene only for illumination. Out of 1,000 sample households, kerosene is used by 732 households. Of these, 169 use kerosene for cooking and for water-heating, 209 use it as the sole cooking fuel, 164 as an auxiliary fuel for cooking and 190 use it for illumination (Table 5.1).

Table 5.1: Pattern of kerosene consumption by Bangalore's households.

PCMI (Rs)		No of HH in the sample	Households using kerosene				
Group	Range		Total	C & WH	C only	Lighting	Auxiliary fuel
I	<100	148	126(17.17)	14(8.28)	15(7.18)	93(48.95)	4 (2.44)
II	100-149	168	147(20.03)	37(21.89)	40(19.14)	59(31.05)	11 (5.49)
III	150-199	112	109(14.85)	29(17.16)	47(22.49)	18 (9.47)	15 (9.15)
IV	200-249	121	112(15.26)	32(18.98)	44(21.05)	14 (7.37)	22(13.41)
V	250-299	120	86(11.72)	23(13.61)	27(12.92)	5 (2.63)	31(18.90)
VI	300-349	138	79(10.76)	21(12.43)	22(10.53)	1 (0.53)	35(21.34)
VII	>350	193	75(10.22)	13(7.69)	14 (6.70)		48(29.27)
	Tota	1000	734(100.0)	169(100.0)	209(100.0)	190(100.0)	166(100.0)

Note: Numbers in parentheses indicate percentages.

If the households using kerosene or both cooking and water-heating are considered, then 58 per cent (98 households) of them belong to the Rs.100-250 per capita monthly income range. Further, the size of the households that use kerosene is small, varying between 3.6 to 4.4, suggesting that kerosene is being consumed mostly by middle- income groups with a small family size. The data also indicate that the per capita consumption of kerosene increases with an increase in income. From 6.7 ltr per capita per month in Group I, the consumption increased to 9 ltr per capita per month in Group VII. In the case of households using kerosene only for cooking, Groups II to IV account for 63 per cent of the households. Here also, the per capita consumption of kerosene increases with an increase in income, from 4.6 to 6.9 ltr per capita per month from Group I to VII (Table 5.2).

The number of households that use kerosene for illumination is 190. Out of these, 152 (80 per cent) belong to Income Groups I and II. Also, the households using kerosene for illumination are from the low-income groups which use cow dung, saw dust

Table 5.2: Per capita consumption of kerosene for cooking and water-heating.

PCMI (Rs)		No.of households using kerosene		Consumption per capita (l/month)		Consumption per household (l/month)	
Group	Range	Cooking and water-Heating	Cooking only	Cooking and water-Heating	Cooking only	Cooking and water-heating	Cooking only
I	<100	14	15	7	5	29.5	23.50
II	100-149	37	40	7	5	27.9	25.00
III	150-199	29	47	7	5	28.4	23.50
IV	200-249	32	44	8	6	30.4	23.10
V	250-299	23	27	8	6	30.5	23.60
VI	300-349	21	22	8	7	31.5	25.20
VII	>350	13	14	9	7	32.4	24.10
Total/average		169	209	7.7	5.7	30.1	24.0

and other agricultural wastes and not kerosene for cooking and water-heating. The quantity of kerosene used does not vary much between these households (2.5 to 3.5 ltr per household per month). This is because most of the households using kerosene for lighting are one-room tenements. During the time of the survey (1984), there were 1,30,000 households in Bangalore without an electric connection and hence, they were forced to depend on kerosene for illumination. Based on the average consumption of 3.1 ltr per household, the consumption of kerosene used for illumination should be about 13 kl per day for 130,000 households.

The dependence of kerosene consumption on income is shown by the kerosene dependence index (KDI) which is the ratio of the percentage of households using kerosene in an income category to the percentage of households belonging to that category. The indices for various income categories (Table 5.3) show that the middle-income groups (II to IV) have a greater dependence on kerosene than other income groups. In the case of kerosene used for illumination, the kerosene dependence index for lighting (KDL) shows that with increasing incomes, the dependence on kerosene for illumination decreases.

Additional support for this conclusion comes from a Lorenz plot in which the cumulative percentage of kerosene-consuming households is plotted against the cumulative percentage of households in various income ranges. The plots drawn for the

Table 5.3: Total kerosene dependence indices for kerosene using households.

PCMI (Rs)		Mid-point of income range (3)	% of HH in the sample (4)	% of HH dependent on charcoal for C&WH (5)	TCDI (col 4/ col 3) (6) = 5/4
Group (1)	Range (2)				
I	<100	50	14.8	8.3	0.6
II	100-149	125	16.8	21.9	1.3
III	150-199	175	11.2	17.2	1.6
IV	200-249	225	12.1	19.0	1.7
V	250-299	275	12.0	13.6	1.1
VI	300-349	325	13.8	12.4	0.9
VII	>350	482	19.3	7.7	0.4

dependence of households on kerosene for cooking and water-heating and also for illumination show that the middle-income groups have a greater dependence on kerosene for cooking and water-heating and the low-income groups for lighting (Figure 5.3).

The kerosene consumption data discussed above show that there is a variation in the consumption of kerosene between different households. This may be due to variations in family income, size of the family and also other energy carriers which are used as supplementaries. Charcoal (CH), LPG and electricity (EL) are used as supplementaries for cooking and firewood (FW). Firewood, charcoal and electricity are used as supplementaries for water-heating. Therefore, it is of interest to determine the impact of these variables on kerosene consumption. Using the regression technique, a model was fitted for kerosene consumption data, for the households that use it for both cooking and water-heating (KCCWH), for those that use it only for cooking (KCC) and for those that use it for lighting. The equations are:

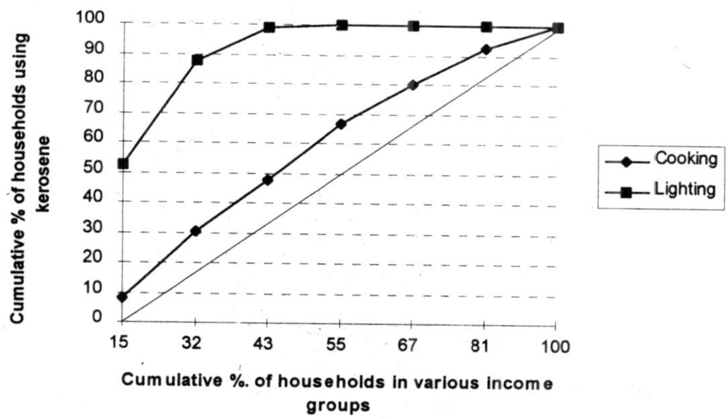

Figure 5.3: Lorenz curve for kerosene usage.

$$KCCWH = \underset{}{12.23} + \underset{(0.015)}{3.86\,FS} + \underset{(0.003)}{0.004\,FI} - \underset{(0.25)}{0.036\,CH} - \underset{(0.001)}{0.023EL} - \underset{(0.02)}{0.02\,LPG} - \underset{(0.009)}{0.008\,FW} \qquad (5.2)$$

$R^2 = 0.695$

$$KCC = 5.32 + \underset{(0.32)}{2.06\,FS} + \underset{(0.0015)}{0.002\,FI} - \underset{(0.001)}{0.0015\,LPG} - \underset{(0.035)}{0.052\,EL} - \underset{(0.001)}{0.015\,CH} \qquad (5.3)$$

$R^2 = 0.65$

$$KCL = 1.37 - \underset{(0.001)}{0.003\,FI} + \underset{(0.16)}{0.93\,NR} \qquad (5.4)$$

$R^2 = 0.54$

In equations 5.2 to 5.4 the values of the correlation coefficient, R_2, indicate that more than 60 per cent of the variation is explained by the variables included in the model. The regression coefficients for *FS* and *FI* show that *FS* is a more important factor than *FI* in determining the variability in kerosene consumption. For example, in the equations, the coefficients for *FI* indicate that kerosene consumption increases by 0.004 and 0.002 units, respectively, for each unit increase in *FI*, whereas kerosene consumption increases by 3.86 and 2.06 units for each unit increase in *FS*. In the case of kerosene used for illumination, the variables included are the family income and number of rooms in the dwelling. Family size is not included because kerosene is used for lighting not on an individual basis but on a joint basis. These two variables explain 54 per cent of the variation in kerosene consumption.

Another procedure for ranking the importance of predictors is to use the step-wise regression analysis in which the variables enter the model according to the value of its R^2, i.e. the explained variation by that particular variable. In equations 5.2 and 5.3 *FS* enters first and is followed by *FI*. For C and WH the supplementary carriers, *EL*, *LPG*, *FW* and *CH* enter in the model in that order and the order for cooking is *LPG*, *EL* and *CH*. The *t* statistic for *FI* is 4 for C and WH, 1.33 for cooking and 3 for lighting with the respective degrees of freedom of 163, 204 and 188. The respective values for *FS* are 15.44 and 6.44. These high values of *t* indicate that the regression coefficients are not zero and there exists a relationship between the variables included in the model (Table 5.4).

Table 5.4: Regression results.

Variable	Degrees of freedom	Mean (l/m)	Coefficient	t value	R^2 (individual)	R^2 (stepwise)
C & WH						
Family size (no.)	168	4	3.860	15.44	0.50	0.5
Family income (Rs/m)	167	685	0.004	4.00	0.14	0.58
Electricity consumption	166	29	0.023	2.59	0.12	0.62
(kWh/m)	165	1	0.020	1.33	0.11	0.65
LPG consumption (kg/m)	164	25	0.008	2.67	0.1	0.68
Firewood consumption (kg/m)	163	1	0.036	1.80	0.08	0.7
Charcoal consumption (kg/m)						
Cooking only						
Family size (no.)	208	5	2.060	6.44	0.50	0.5
Family income (Rs/m)	207	957	0.002	1.33	0.14	0.61
Electricity consumption	206	1	0.015	15.00	0.09	0.63
(kWh/m)	205	28	0.052	1.44	0.05	0.64
LPG consumption (kg/m)	204	1	0.015	7.50	0.03	0.65
Firewood consumption (kg/m)						
Lighting only						
Family income (Rs/m)	189	170	0.003	3.00	0.43	0.43
Rooms (no.)	188	1	0.930	5.81	0.25	0.54

Table 5.5: Monthly expenditure on kerosene for various end-uses.

PCMI (Rs)		Expenditure on kerosene (Rs/cap/m)			Expenditure as % of income		
Group	Range	Cooking and water-heating	Cook-ing	Lighting	Cooking and water-heating	Cook-ing	Lighting
I	<100	16.1	11.0	8.0	21.5	14.7	10.7
II	100-149	16.3	12.0	8.0	13.0	9.6	6.4
III	150-199	17.0	12.0	10.0	9.7	6.9	5.7
IV	200-249	18.2	13.2	9.0	8.1	5.9	4.0
V	250-299	19.7	14.9	11.0	7.2	5.4	4.0
VI	300-349	20.0	16.3	11.0	6.2	5.0	3.4
VII	>350	21.0	16.6		4.2	3.3	0.0
Average		18.3	13.7	9.5	10.0	7.2	4.9

The monthly expenditure on kerosene for both cooking and water-heating and as the sole cooking fuel increases with an increase in income. But if the expenditure is considered as a percentage of income, then it is a decreasing percentage. In Group I, households spend 20.8 per cent of their income on cooking and water-heating and in Group VII, it is 5.4 per cent. In the case of cooking only, the percentages of income spent on kerosene in Group I and Group VII are 14.72 and 4.14, respectively. For lighting the percentage of income spent on kerosene is 10 per cent in Group I and 3.23 per cent in Group VI (Table 5.5).

Industrial consumption

Permits for selling superior kerosene oil are being issued to industries for purposes other than cooking and illumination. The requirement of kerosene in an industry is assessed by the metallurgist of the Industries and Commerce Department. He gives an essentiality certificate recommending the quantity of kerosene to be issued on the basis of which the Directorate of Food and Civil Supplies issues the permit. There were 287 large- and small-scale industries consuming kerosene. During the period 1980-90, the consumption of kerosene rose from 4,200 to 12,100 kl per day (Figure 5.4). The number of industries falling into each category and the quantity of kerosene consumed by each industry are shown in Table 5.6.

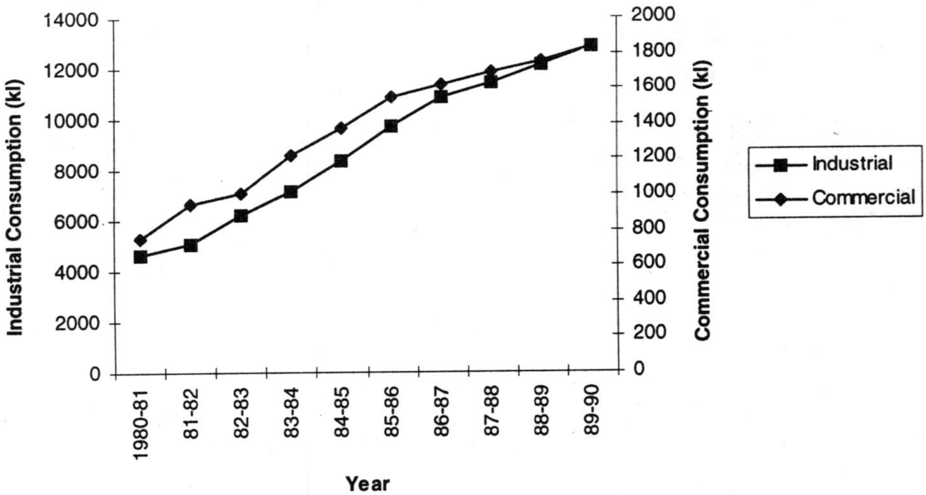

Figure 5.4: Industrial and commercial consumption of kerosene.

Out of a total monthly consumption of 680 kl, 33.8 per cent is consumed by 9 industries in the consumer category of >10 kl and 48.8 per cent is consumed by small industries, each with a per capita monthly consumption of <1.0 kl. The industries were divided into four groups on the basis of their monthly consumption. The main end-uses in industries are: (i) process heating, (ii) as a coolant, (iii) overhauling and (iv) cleaning. Cleaning, with 56 per cent of the total consumption, is the major end-use in industry. Other end-uses like process-heat, overhauling and coolant share equal percentages (around 12 per cent).

Table 5.6: Pattern of kerosene consumption in industries.

Sl.no	Consumption level (kl/m)	Industries (no.)	Consumption (kl/m)	% of the total
1	<1.0	253	332	48.82
2	1.0-4.99	15	30	4.41
3	5.0-9.99	10	88	12.94
4	>10	9	230	33.82
Total		287	680	100.00

Commercial consumption (restaurants)

The Directorate of Food and Civil Supplies issues kerosene permits to restaurants just to prevent them from cornering the supply meant for domestic consumption *via* retailers and hawkers. The quantity of kerosene consumed by restaurants increased from 752 kl in 1980-81 to 1,794 kl in 1989-90, an increase of 150 per cent. The number of restaurants obtaining kerosene during 1980-81 was 390 with a consumption of 5.4 litres per day per establishment. The number of restaurants increased to 550 in 1983-84 and the consumption per establishment was 6.9 litres per day. This increase in per capita consumption may be due to the substantial increase in the prices of other fuels such as firewood and electricity.

Other types of consumers

Hostels: The number of hostels that were getting kerosenewas 60. Their total consumption was 0.5 kl per day which works out to 8.3 ltr per hostel per day.

Defence establishments: The 32 Defence establishments in Bangalore use kerosene in their canteens for cooking and also use it for overhauling their vehicles. On an average, these establishments consume one kl per day.

Karnataka State Road Transport Corporation: Karnataka State Road Transport Corporation (KSRTC) uses kerosene mainly for the purpose of overhauling vehicles. On an average, the 13 bus depots in Bangalore consume 3 kl per day.

Road asphalting agencies: The Public Works Department of the Karnataka State Government uses kerosene for heating tar. On an average, 3.5 kl of kerosene per day is consumed for asphalting Bangalore's roads.

Besides the above categories, other consumers are: (i) government press, (ii) newspaper offices, (iii) film studios, (iv) All India Radio, and (v) coffee and tea stalls. These customers together consume around 10 kl per day.

Consumer-wise break-up

Households are the largest consumers of kerosene with an average daily consumption of 280 kl accounting for 86.15 per cent of the total consumption. The next major consumer is industry with a daily consumption of 23 kl (7.1 per cent of total). The remaining consumption is accounted for by restaurants, hostels, etc.

End-use analysis

Cooking is the major end-use of kerosene (70.8 per cent) with a daily consumption of 230 kl followed by water-heating with 54 kl (16.6 per cent). The remaining kerosene is used for other end-uses such as illumination, process-heating, cleaning, etc (Figure 5.5).

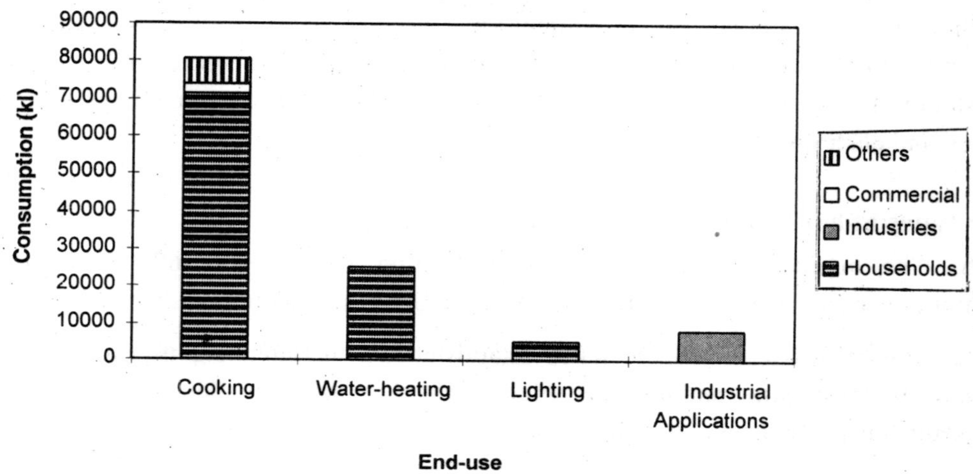

Figure 5.5: End-use analysis of kerosene in industries.

Discussion

(i) Impact on the transportation system

Kerosene is transported by railway wagons and trucks, their shares being 90 per cent and 10 per cent, respectively. On an average, Bangalore gets its freight through 630 trucks and 125 wagons per day. Out of this, kerosene's share is 1 per cent of the total in the case of trucks and 1.5 per cent in the case of wagons.

(ii) Energy cost of kerosene supply

Bangalore consumes about 300 kl of kerosene per day which is brought from Madras situated 350 km away. This corresponds to about 105,000 kl km per day for wagons. The total quantity of diesel used to transport the kerosene is 1,278 ltr per day.

The energy content of the average daily consumption of 1,278 ltr of diesel is 47.29 x 10^9 joules (J) (calorific value per ltr of diesel = 37.00 x 10^6 J). This energy is used to transport 300 kl or 10.5 x 10^{12} J of kerosene energy (calorific value per litre of kerosene = 35.00 x 10^6 J). But the efficiency with which it is being used (for cooking and water-heating) is 45 per cent of the primary energy which amounts to 1.9 x 10^{12} J. Therefore, to transport 10.5 x 10^{12} J of energy, 21.6 x 10^9 J is used in the process (here, transport from drilling to refinery is not considered).

Liquefied Petroleum Gas (LPG)

The consumption of LPG is increasing rapidly. It was 9,268 metric tons during 1980-81 which increased to 30,656 metric tons during 1989-90, an almost three-fold increase in 10 years (Table 5.7).

Table 5.7: Quantity of LPG consumed (in tonnes) in Bangalore (1980-90).

Month	1980-81	81-82	82-83	83-84	84-85	85-86	86-87	87-88	88-89	89-90
Consumption (tonnes)	9268	10438	12617	16051	19092	20785	22659	25206	27390	30656
Avg/month	772	870	1051	1337	1591	1732	1888	2100	2282	2554
Std.dvn	136	107	100	167	96	116	127	121	89	123
Variation coefficient (%)	17.7	12.3	9.5	12.5	6.0	6.7	6.8	5.8	3.9	4.8
Average/day	25	29	35	44.0	52	56.9	62.1	69.1	75.0	84.0

The following equation shows the trend in the LPG consumption (LPGC). The extremely high value of the coefficient of correlation, R^2, shows that the increase of consumption is linear,

$$LPGC = 6288 + Y\,2383 \; (Y = \text{Year}) \quad (5.5)$$
$$(260.2)$$

$$R^2 = 0.99$$

Residential Consumption

The residential sector with 75 per cent of the total consumption is the major consumer of LPG. Its consumption is increasing at a constant rate of approximately 15 per cent per annum. The yearly consumption which stood at 6,785 metric tons during 1980-81 increased to 25,100 metric tons during 1989-90, a nearly four-fold increase in ten years. There is also an increase in the number of LPG consumers, the number increasing from 68,400 to 1,48,000, a 2.4 times increase from 1980 to 1990. The increase in the magnitude of residential consumption is not proportional to the increase in the number of consumers per household consumption increased from 7.91 to 9.15 kg per month (Figure 5.6).

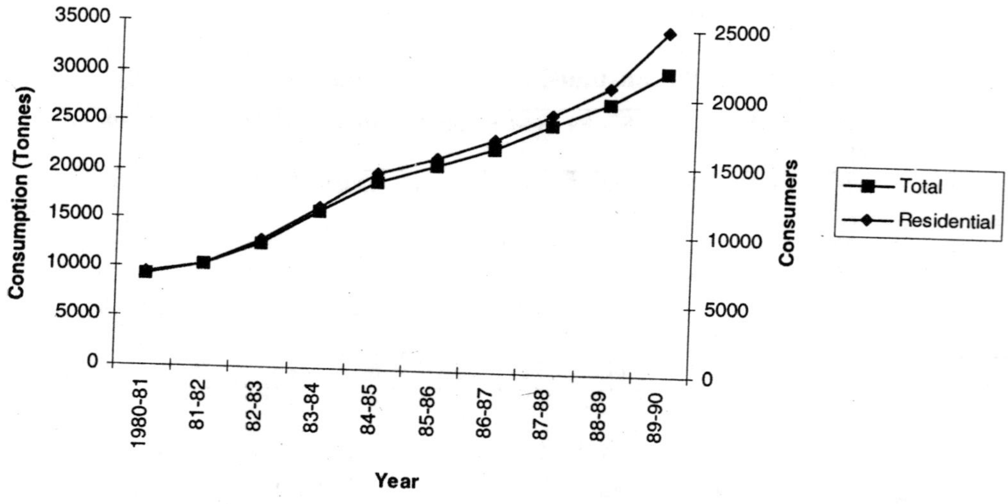

Figure 5.6: Residential LPG consumption (1980-90).

The household survey data were studied to determine the relationship between LPG consumption and other socio-economic factors that influence consumption. The average per capita consumption per month, according to the survey, was 5.5 kg per household for cooking and water-heating and 4 kg for cooking only. These averages need to be dissected and the variations in different income groups found. Hence, the study looked closely at the relationship between the use of LPG and the family's income.

Table 5.8: Pattern of LPG consumption in Bangalore's households.

Per capita monthly income (Rs) Group	Range	No of households In the sample	Number and percentage of households using LPG Total	Cooking & Water-heating	Cooking only	Auxiliary
I	<100	148	0	0	0	0
II	100-149	168	0	0	0	0
III	150-199	112	3 (1.33)	2 (4.76)	1 (0.86)	0
IV	200-249	121	9 (4.00)	3 (7.14)	4 (3.45)	2 (2.99)
V	250-299	120	26 (11.56)	6 (14.29)	11 (9.48)	9 (13.43)
VI	300-349	138	53 (23.56)	8 (19.05)	19 (16.38)	26 (38.81)
VII	>350	193	134 (59.56)	23 (54.76)	81 (69.83)	30 (44.78)
Total		1000	225 (100.0)	42 (100.0)	116(100.0)	67 (100.0)

Note: Numbers in parentheses represent percentages.

Families with different incomes have markedly different patterns of energy usage as might be expected. The number of households using LPG in different income groups varies considerably (Table 5.8).

In groups I and II, the consumption of LPG does not figure at all. From then, the number of households using LPG rises with increase in income. For example, in group III, 3.66 per cent of the households use LPG whereas in group VII almost 75 per cent use it. Therefore, it is only in high income groups that LPG can be considered to be a dominant energy source for cooking; it is not used at all by the low-income groups. This is due to the high initial cost of the stove (Rs1,200) and the deposit for the connection (Rs 500). LPG is preferred by high-income groups because it is a clean, efficient and convenient fuel. There are three types of households that use LPG. The first type consists of households which use it for both cooking and water-heating, the second type depends on it entirely for cooking and the last group uses it as a supplementary fuel for cooking along with either kerosene or electricity. Of the 225 households using LPG, 42 fall in to the first category, 116 in the second and 67 into third category.

To study the dependence of consumption on income, the total LPG dependence index, LPGDI, for various income groups is presented in Table 5.9. It can be seen that the percentage of households with a per capita monthly income greater than Rs. 300 depending on LPG is greater than their percentages in the sample.

Another way of showing the income dependence of LPG consumption is by drawing the Lorenz curve (cumulative percentages of LPG using households *vs*

Table 5.9: LPG dependence indices of households.

PCMI (Rs)		Mid-point of income range	% of households in the category	% of households using LPG	LPGDI
Group	Range				
I	<100	-	14.8	0.0	0.0
II	100-149	-	16.8	0.0	0.0
III	150-199	180	11.2	1.3	0.1
IV	200-249	226	12.1	4.0	0.3
V	250-299	273	12.0	11.5	1.0
VI	300-349	323	13.8	23.6	1.7
VII	>350	1068	19.3	59.5	3.1
	Total		100.0	100.0	

cumulative percentages of households in different income groups). The curve shows a greater concentration of LPG-dependent households in high-income groups (Fig 5.7).

The per capita consumption of LPG varies positively with income in all the three types of households. But, the consumption rises more slowly than income if we go from one economic group to another. The per capita consumption in Group III is 3.64 kg per month for cooking and water-heating and 3.31 kg for cooking alone. The corresponding

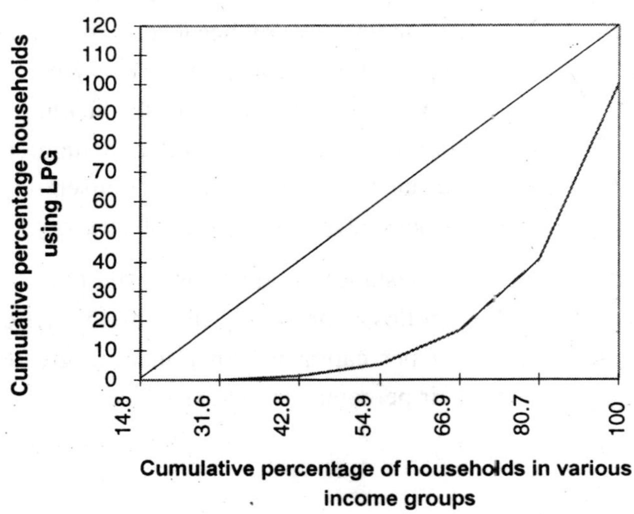

Figure 5.7: Lorenz Curve for LPG usage.

Table 5.10: Per capita consumption of LPG in different income groups (cooking and water-heating).

PCMI (Rs)		No of households using LPG		Consumption (kg/m)			
				Per capita		Per household	
Group	Range	Cooking and water-heating	Cooking only	Cooking and water-heating	Cooking only	Cooking and water-heating	Cooking only
I	<100	-	-	-	-	-	-
II	100-149	-	-	-	-	-	-
III	150-199	2	1	4	3	15.47	14.90
IV	200-249	3	4	5	4	19.52	15.64
V	250-299	6	11	5	3	17.55	14.68
VI	300-359	8	19	6	4	19.62	16.45
VII	>350	23	81	5.7	4.22	18.34	18.91
	Total	42	116				

figures for Group VII are 5.70 and 4.22 kg (Table 5.10).

The table shows that with an increase in income, households tend to consume more fuel probably due to the increase in the number of dishes prepared. It was observed that with an increase in the family size, the quantity of fuel used increases. Many of the households using LPG depend, to a certain extent, on supplementary energy carriers like kerosene (KR), charcoal (CH), firewood (FW) and electricity (EL). Therefore, to determine the effect of family income (FI) (Rs/month), family size (FS) (number) and the consumption of supplementary fuels on household LPG consumption (LPGC) (kg/month), the regression method was used.

To study the impact of various household parameters on LPG consumption, regression analysis was carried out. LPG consumption was regressed on the family income, family size and consumption of supplementary fuels to yield the following results:

For cooking and water-heating:

$$\begin{aligned} LPGC &= 4.32 + \underset{(0.413)}{5.205\,FS} + \underset{(0.002)}{0.003\,FI} - \underset{(0.093)}{0.007\,FW} \\ &\quad - \underset{(0.19)}{0.08\,CH} - \underset{(0.01)}{0.01\,EL} \end{aligned} \qquad (5.6)$$

$R^2 = 0.82$

For cooking only:

$$LPGC \;=\; \underset{}{5.32} + \underset{(0.45)}{3.4\,FS} + \underset{(0.0001)}{0.002\,FI} - \underset{(0.78)}{1.2\,KR} \qquad (5.7)$$

$R^2 = 0.84$

(The figures in parentheses refer to the standard errors of the coefficients.)

The equations show that the variables included in the model explain more than 80 per cent of the variation in the LPG consumption. Also, the individual regression coefficients in both the equations show that *FS* seems to be the major determinant of LPG consumption. The values of the *t* statistic show that the results are significant at 99 per cent confidence level (Table 5.11).

The stepwise regression results shown in the table indicate that family size is the major determinant that influences LPG consumption.

The per capita consumption of LPG increases with income as has been discussed earlier. But, the percentage of income spent on LPG decreases with increase in income (Table 5.12).

In Group III, households that use LPG for both cooking and water-heating sp·

Table 5.11: Regression results.

Variable	Mean	Coefficient	*t* value	R^2 (individual)	R^2 Stepwise
C&WH					
Family size (no.)	4.0	5.12	11.91	0.68	0.68
Family income (Rs/month)	1870.0	0.002	2.00	0.33	0.75
Electricity consumption	18.5	0.01	1.00	0.08	0.80
(kWh/m)	2.5	0.92	12.00	0.07	0.83
Kerosene consumption (l/m)	25.0	0.006	0.75	0.05	0.84
Firewood consumption (kg/m)	4.0	0.07	3.50	0.04	0.85
Charcoal consumption (kg/m)					
Cooking only					
Family size (no.)	4.4	3.4	7.56	0.70	0.70
Family income (Rs/month)	3089.0	0.002	20.00	0.16	0.83
Kerosene consumption (l/m)	1.6	1.2	1.54	0.04	0.84
Electricity consumption (kWh/m)	25.0	0.05	17.00	0.001	1.00

Table 5.12: Expenditure on LPG consumption.

PCMI (Rs)		No of HH in the group	Expenditure (RS/cap/m)		Expenditure as % of income	
Group	Range		Cooking and Water-heating	Cooking only	Cooking and water-heating	Cooking only
I	<100	148	-	-	-	-
II	100-149	168	16.67	15.16	9.42	8.42
III	150-199	112	21.43	16.85	9.32	7.46
IV	200-249	121	24.73	15.89	8.96	5.82
V	250-299	120	25.24	17.36	7.81	5.37
VI	300-359	138	26.24	19.33	5.21	1.81
VII	>350	193				
	Total	1000				

9.42 per cent of their incomes on this fuel, while those that use it only for cooking spend 8.42 per cent of their incomes; in Group VII, the corresponding figures are 5.21 and 1.81 per cent. This means that fuel, like other necessities such as food and housing, claims a larger share of the budget of middle-income groups than that of the affluent groups.

Industrial and commercial consumption

Permits are being given by Oil Corporations to industries to use LPG for purposes other than cooking. Special permits are also being issued to industrial canteens. The quantity increased from 101.9 metric tons to 176.5 metric tons per month (an increase of 75 per cent) from 1980 to 1990. Of the total industrial consumption, nearly 80 per cent is used for process heat and the remaining for cooking in the industrial canteens. The commercial sector includes restaurants, hostels, canteens, offices and laboratories. The daily consumption during 1980-81 was 3.4 metric tons per day which increased to 11.8 metric tons during 1986-87, an increase of 250 per cent. There are four types of restaurants (A, B, C, and D) in Bangalore. They are classified based on turnover. A sample survey covering 5 per cent of each type indicates that there is no consumption of LPG in the D type of restaurants. In the C type, LPG consumption is marginal. On an average, five cylinders (each cylinder carries 15 kg) are consumed per month by these restaurants. In the A and B types the consumption varied between 30 and 50 cylinders (450-750 kg) per month. Figure 5.8 shows the quantity of LPG consumed by the industries and commercial establishments over a period of eight years.

The hostels in Bangalore can be classified into three types depending on whether they are associated with (i) colleges and working people, (ii) schools and convents and (iii) aided hostels. LPG is used in the first and second categories but its consumption is less when compared with other fuels. The 120 hostels in Bangalore consume, on an average, 10 to 20 cylinders per hostel per month. In hostels, LPG is mainly used for cooking.

In addition to restaurants and hostels, LPG is consumed in hospitals and laboratories mainly for process heat. The consumption per lab/hospital varies from 1 to 20 cylinders per month. On an average, the consumption by the laboratories and hospitals is 10 cylinders per day.

Sector-wise consumption

The residential sector is the major consumer of LPG with 75 per cent of the total consumption. Its share in the total consumption increased only marginally (two per cent) between 1980 and 1990. The shares of the industrial and commercial sectors are 10 and 15.5 per cent, respectively. The remaining quantity is consumed by other consumers like hospitals, laboratories, etc. The consumption by industries decreased from 13 to 8 per cent of the total and commercial consumption increased from 13 to 17 per cent between 1980 and 1990 (Table 5.13).

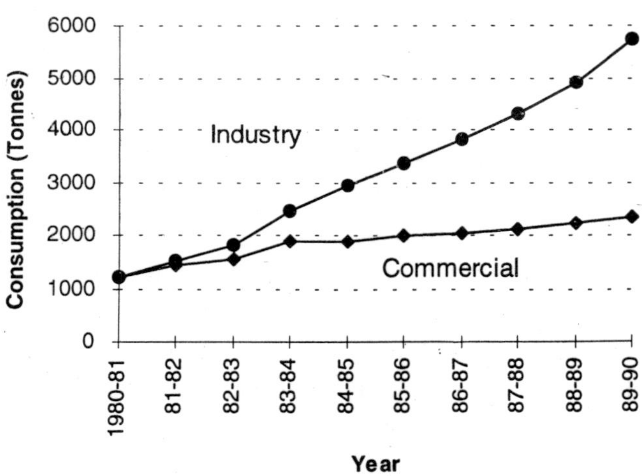

Figure 5.8: Commercial and industrial LPG consumption.

Table 5.13: Sectoral Consumption.

Year	Sectoral consumption (tonnes)			
	Residential	Commercial	Industrial	Others
1980-81	6785 (73.21)	1224 (13.21)	1222 (13.19)	37 (0.40)
1981-82	7441 (71.29)	1499 (14.36)	1433 (13.81)	65 (0.62)
1982-83	9239 (73.23)	1817 (14.40)	1537 (12.21)	24 (0.19)
1983-84	11690 (72.83)	2449 (15.26)	1878 (11.73)	34 (0.21)
1984-85	14183 (74.29)	2959 (15.50)	1895 (9.95)	55 (0.29)
1985-86	15369 (73.9)	3348 (16.10)	1986 (9.6)	82 (0.39)
1986-87	16766 (74.0)	3810 (16.80)	2024 (8.9)	59 (0.26)
1987-88	18174 (74.2)	4311 (17.10)	2118 (8.4)	63 (0.25)
1988-89	20625 (74.1)	4932 (16.91)	2249 (9.0)	69 (0.24)
1989-90	23420 (75.9)	54125(17.02)	2402 (8.9)	76 (0.25)

Note: Numbers in parentheses indicate percentages.

End-use analysis

As already mentioned, LPG is used for cooking, water-heating and process heating. Cooking is the major end-use of LPG with a daily consumption of 43 metric tonnes (82.7 per cent of the total). The next major end-use is water-heating with a consumption of 5 metric tons per day (9.7 per cent). The remaining quantity is used for process heating (Figure 5.9).

Discussion

Impact on the transport system

About eight trucks a day transport LPG to the city which means that nearly 1.3 per cent of the 630 truck traffic is committed to LPG transport.

Energy cost of transporting LPG

The transport of 50 metric tonnes of LPG or 2.28 x 10^{12} joules of energy is through trucks which consume diesel. The distances from Bombay and Koyali (where the LPG refineries are located) to Bangalore are 1,000 and 1,200 km., respectively. Thus the total energy consumption through these trucks comes to about 52,000 tonne km/day which requires 630 litres of diesel per day.

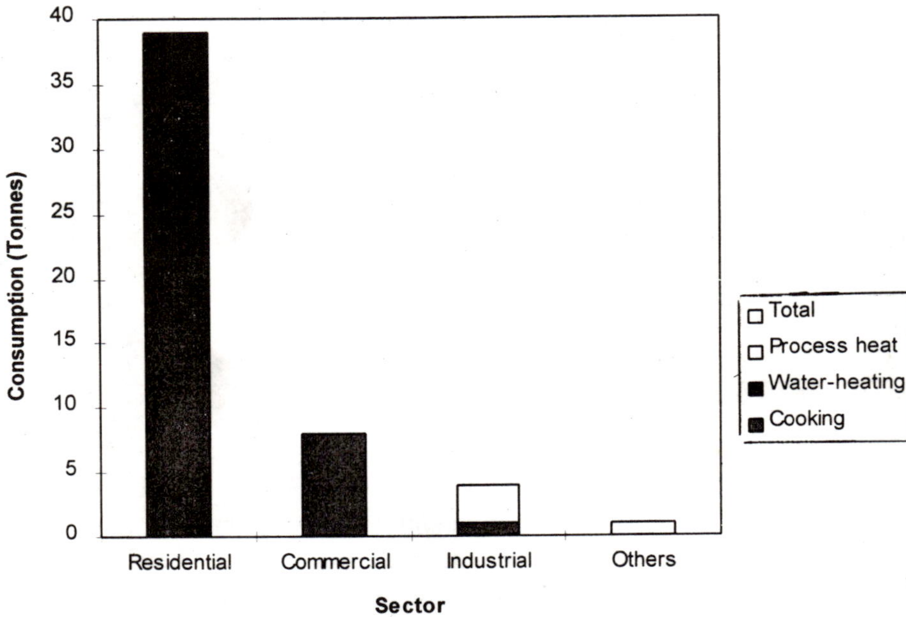

Figure 5.9: End-use analysis of LPG.

Supply and safety constraints

Though LPG is very popular with housewives, its irregular supply often causes inconvenience, thus forcing them to resort to other fuels like electricity and kerosene. Reports of accidents with LPG have also deterred potential users. Some safety device to detect leakages and quick disconnection of its supply in case of an accident would make LPG more popular.

Other Petroleum Products

The main petroleum products considered here are (i) motor spirit (MS), (ii) high speed diesel (HSD), (iii) light diesel oil (LDO) and (iv) furnace oil (FO).

Consumption

Figure 5.10 depicts the trend in the consumption of various petroleum products over a period of 10 years, i.e. 1980 - 90.

Motor spirit is commonly referred to in India as petrol. It is used mainly for automobile engines. From 1980-81 to 1989-90, the consumption of petrol increased from 58,088 to 187,415 kl. High speed diesel (HSD) is used in stationary engines for generating power and automobile engines. It is used in slow-speed engines (<500 rpm) of stationary units, medium speed engines (500-1500 rpm) of tractors, locomotives and pumps, and high-speed engines (>1500 rpm) of trucks, buses, etc. The consumers of HSD can be classified into the following categories: (i) railways, (ii) industries, (iii) automobiles (through retail depots), (iv) Karnataka State Road Transport Corporation (KSRTC), and (v) defence establishments. The consumption of HSD has been increasing rapidly over the years, from a figure of 125,265 kl during 1980-81 to 296,927 kl during 1989-90, a two-fold increase. Of all the users, the share of vehicles (through retail) is the highest, followed by industries and railways. Light diesel oil is used mainly in stationary engines in industries for captive power and in furnaces for process heat. Its consumption in recent years has increased from 11,308 kl in 1980-81 to 18,673 kl in

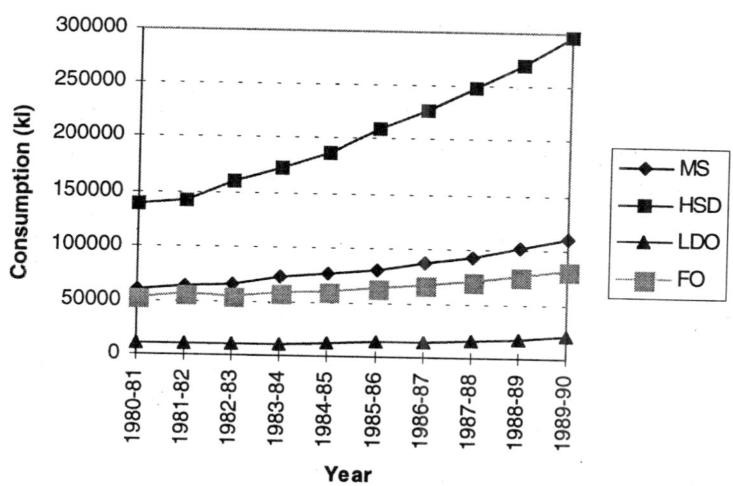

Figure 5.10: Consumption of various petroleum products.

manufacturing industries to generate steam for furnaces. Its consumption has increased considerably in recent years, from 52,984 in 1980-81 to 79,641 kl in 1989-90, an increase of 4.3 per cent per annum (Table 5.14).

The following equations describe the annual increase for different petroleum products:

$$MS = 53816 + 3491.4\ Y \text{ (Y = Year)} \quad (5.8)$$
$$(485.4)$$
$$R^2 = 0.95$$

$$HSD = 112862 + 11457.6\ Y \quad (5.9)$$
$$(1219.7)$$
$$R^2 = 0.97$$

$$LDO = 9369 + 738\ Y \quad (5.10)$$
$$(80.9)$$
$$R^2 = 0.98$$

$$FO = 49136 + 2553\ Y \quad (5.11)$$
$$(543.7)$$

Table 5.14: Consumption of petroleum products (1980-90).

Product	Sector	Consumption (kl)									
		1980 -81	1981 - 82	1982 - 83	1983 - 84	1984 - 85	1985 - 86	1986 - 87	1987- 88	1988- 89	1989- 90
MS	Retail	56945	59174	62963	67285	71022	74860	80883	86324	94372	103146
	Industry	2875	2770	1978	3247	3629	3851	3927	4192	4237	4472
	Defence	1196	1236	1289	1406	1542	1715	1922	2240	2371	2465
	Total	61016	63180	66230	71938	76193	80426	86732	92756	100980	110083
HSD	Retail	78215	75764	85029	92265	103481	121680	134538	150235	167128	187415
	Industry	26370	28241	33550	35789	36652	38117	40282	42174	44390	47245
	KSRTC	23868	25710	26970	27154	27420	28736	29117	30127	31445	32892
	Railways	8475	10341	12671	15290	17141	19342	21465	23451	24892	26975
	Defence	1216	1304	1346	1528	1672	1826	1930	2106	2285	2400
	Total	138144	141360	159566	172026	186366	209701	227332	248093	270140	296927
LDO	Industry	10802	10755	10528	10876	11236	13161	14260	15451	16204	17650
	Dealer	506	458	461	470	501	568	712	805	912	1023
	Total	11308	11213	10989	11346	11737	13729	14972	16256	17116	18673
Fuel oil		52984	56600	54142	58236	60265	63927	67521	71342	75490	79641

$$FO = 49136 + \underset{(543.7)}{2553\,Y} \qquad (5.11)$$

$R^2 = 0.79$

(The numbers in parentheses are the standard errors.)

End-use analysis

Petroleum products are used for transport, power generation and process heating. During 1983-84, for instance, Bangalore consumed 15,520 TJ in the form of petroleum products out of which HSD's contribution was the highest (54.2 per cent) followed by MS and FO with 23.3 per cent and 18.8 per cent, respectively. This energy was utilised for transportation (46.7 per cent), in industry for power generation (11.5 per cent), and for process heat (15.5 per cent). The remaing was used by ;idential and commercial sectors. In the transport services, diesel, which is used by railways, buses and trucks, etc., is the main fuel accounting for two-thirds of the total while for the rest it is petrol.

Bettween 1980 and 1990, Bangalore exhibited fairly stable transport energy shares, growing only from 46.1 to 49.03 per cent. The variation of different energy carriers in different sectors indicates that there is a small decrease in the share of industrial energy consumption while the residential and commercial sectors' share remained constant. In the total consumption, transportation heads the list with about 45 per cent of the total followed by the industrial sector with 25 per cent. The remaining quantity is consumed in the residential and commercial sectors. Sector-wise trends over time in the consumption of energy show transport as the area which has an increasing demand (Table 5.15).

Petroleum products and transportation

Modern transport sectors in most developing countries show another predominant trend – the major share is held by road transport. This was 64.7 per cent during 1980-81 and increased to 91.28 per cent in 1989-90, whereas the railways show a corresponding decrease (Table 5.16). The main reason for this may be the shift from coal to oil which is more efficient. The breakdown of fuel type by mode of transport provides an interesting glimpse of the variability in the transport energy use pattern.

Table 5.15: Consumption of petroleum products by the end-use sector.

Year	Total consum-ption (TJ)	Transportation				Industry						Residential and commercial	
		Motor spirit	Diesel	Total	As % of total	Diesel	Fuel oil	Light diesel oil	Kero-sene	Total	As % of total	Kero-sene	As % of total
1980-81	13291	89	4136	6125	46.1	976	2230	393	163	3761	28.3	3406	25.6
1981-82	13757	2060	4185	6245	45.4	1045	2133	384	178	3741	27.2	3772	27.4
1982-83	14908	2159	4663	6822	45.8	1241	2294	397	218	4151	27.8	3935	26.4
1983-84	15912	2345	5042	7387	46.4	1324	2374	410	251	4359	27.4	4166	26.2
1984-85	17186	2484	5539	8023	46.7	1356	2519	480	292	4647	27.0	4515	26.3
1985-86	18827	2622	6346	8968	47.6	1410	2660	520	340	4931	26.2	4928	26.2
1986-87	20551	2827	6921	9748	47.4	1490	2811	564	380	5245	25.5	5558	27.0
1987-88	22301	3024	7619	10643	47.7	1560	2970	588	399	5518	24.8	6139	27.5
1988-89	24227	3292	8353	11645	48.07	1642	3168	618	440	5870	24.2	6712	27.7
1989-90	26161	3589	9238	12827	49.03	1748	3401	625	460	6234	23.8	7109	27.1

Source: Primary data from Indian Oil Corporation, Bharat Petroleum Corporation, Hindustan Petroleum Corporation, and Indo-Burman Petroleum Corporation.

The first trend of interest in the end-use analysis of petroleum products is that 66 per cent of the total consumption goes to the transport services and the rest for industrial purposes such as process-heating and power generation. Here, diesel, which is used by railways, buses, trucks, etc., is the main fuel accounting for 59.32 per cent of the total while the needs of other types vehicles like cars and two- and three-wheelers are met by petrol. A study of the variation of the number of different vehicles over time (i.e. from

Table 5.16: Consumption of energy by mode of transport (1980-90).

Year	Total consumption (TJ)	Road transport				Railway transport			
		Motor Spirit	Diesel	Total	As % of total	Diesel	Coal	Total	As % of total
1980-81	9068	1989	3877	5866	64.70	258	2943	3201	35.30
1981-82	8746	2060	3872	5931	67.82	314	2501	2814	32.18
1982-83	8921	2159	4280	6439	72.18	383	2099	2482	27.82
1983-84	9058	2345	4573	6918	76.38	469	1671	2140	23.62
1984-85	9167	2484	4974	7458	81.35	566	1144	1710	18.65
1985-86	9870	2622	5712	8334	84.44	634	902	1536	15.56
1986-87	10345	2827	6205	9033	87.31	716	597	1313	12.69
1987-88	10994	3024	6825	9849	89.58	794	351	1145	10.42
1988-89	11835	3292	7440	10732	90.68	913	190	1103	9.32
1989-90	12907	3589	8192	11781	91.28	1046	80	1126	8.72

1980-81 to 1989-90) shows that two-wheelers and cars increased steadily with time while the increase in other types of vehicles was marginal (Figure 5.11).

During the period 1980-90, the average growth rate of vehicular population was 13.22 per cent. The variation of different categories of vehicles between 1980–90 indicates that, during 1980-81, two-wheelers constituted 64.63 per cent of the total and increased to 76.75 per cent during 1989-80. All other categories showed a declining share in the total (Anon, 1991). The growth of two-wheelers is remarkable after 1980. They achieved an average growth rate of 15.41 per cent per year during this period, certainly the highest among the developing countries. Although the growth rate of cars was slow (9.75 per cent) when compared to two-wheelers, the total number of cars increased steadily over the years. The growth rate of taxies was 9.7 per cent. During the same period, the growth rate of petrol consumption was 6.79 per cent (Table 5.17). In the case of diesel-driven vehicles (Table 5.18), trucks achieved an average of 8.18 per cent growth per annum while that of buses was only 5.11 per cent. During this decade, diesel consumption reported an average growth rate of 8.15 per cent per year.

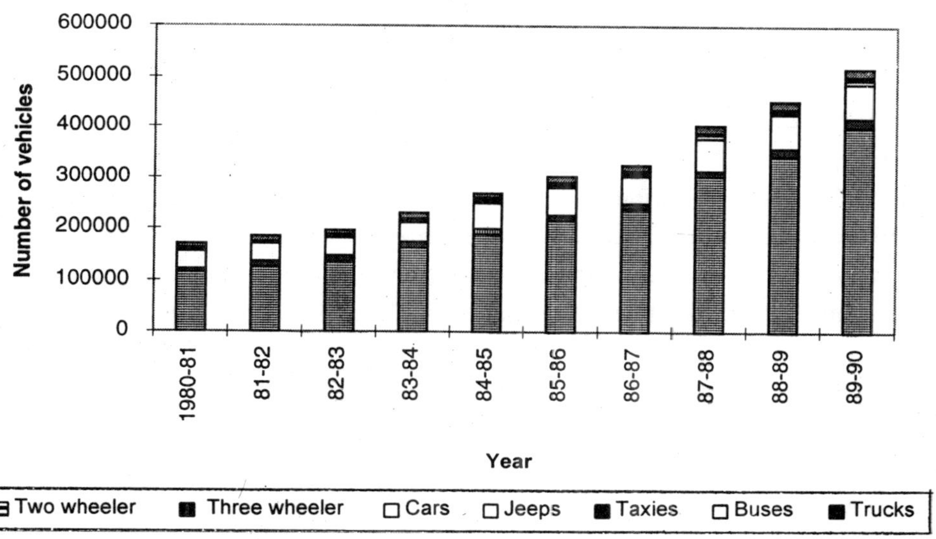

Figure 5.11: Growth of various types of vehicles.

Table 5.17: Growth rate of motor vehicles and petrol consumption (1980-90).

Year	Vehicle type and its growth rate (%)											
	Two wheelers	Growth rate (%)	Three-wheelers	Growth rate (%)	Cars	Growth rate (%)	Jeeps	Growth rate (%)	Taxies	Growth rate (%)	Consumption (kl)	Growth rate (%)
1980-81	111750		10046		31738		3554		1120		61016	
1981-82	125600	12.4	10355	3.1	32429	2.2	3570	0.5	1171	4.6	63180	3.6
1982-83	135536	7.9	11063	6.8	33506	3.3	3726	4.4	1276	9.0	66230	4.8
1983-84	162206	19.7	11880	7.4	36064	7.6	4043	8.5	1607	25.9	71938	8.6
1984-85	189619	16.9	11348	-4.5	47787	32.5	4995	23.6	2027	26.1	76193	5.9
1985-86	216234	14.0	10446	-7.9	53150	11.2	4914	-1.6	1756	-13.4	80426	5.7
1986-87	236726	9.5	10524	0.7	54885	3.3	5122	4.2	1717	-2.2	86732	7.8
1987-88	301677	27.4	12439	18.2	64703	17.9	5741	12.1	2027	18.1	92756	6.9
1988-89	343733	13.9	15513	24.7	65916	1.9	5966	3.9	2089	3.1	100980	8.9
1989-90	401811	16.9	15591	0.5	71078	7.8	5823	-2.4	2218	6.2	11083	9.0
Average growth rate		15.4		5.4		9.8		5.9		8.6		6.2

The above results indicate that the main characteristics of the Bangalore transportation system is the dominance of personal passenger vehicles and the consumption of petrol for these vehicles. Passenger transportation dominates freight in energy use and personal vehicles dominate passenger transport. The breakdown of fuel type by mode of transport provides an interesting glimpse of the variability in the transport energy use pattern.

Vehicle type and energy use

Vehicle census approach

The total petrol and diesel consumption discussed till now does not reveal the relationship between the various categories of vehicles and the total consumption, which is, therefore, an important parameter to determine. Instead of finding these values by sample surveys, the vehicle-wise consumption can be estimated using statistical methods. This method can be referred to as the "vehicle census approach". Using this approach, the fuel consumed by different vehicles was estimated by regressing the average monthly fuel consumption on the number of vehicles in different categories. The form of the equation to estimate the monthly fuel consumption of *i*th category of vehicles is:

Table 5.18: Growth rate of motor vehicles and diesel consumption (1980-90).

Year	Vehicle type and its growth rate (%)									
	Buses	Growth rate	Trucks	Growth rate	Tractors	Growth rate	Trailors	Growth rate	Diesel (kl/year)	Growth rate (%)
1980-81	2256		8236		1929		1734		103299	
1981-82	2239	-0.7	8545	3.7	1781	-7.7	1637	-5.6	102778	-0.5
1982-83	2482	10.9	10020	17.2	1823	2.4	1678	2.5	113345	10.3
1983-84	2704	8.9	10686	6.6	1510	-17.2	1361	-18.9	120763	6.5
1984-85	2926	8.2	11106	3.9	1493	-1.1	1444	6.1	132573	9.8
1985-86	2746	-6.1	11532	3.8	755	-49.4	671	-53.5	152242	14.8
1986-87	3030	10.3	11366	-1.4	496	-34.3	446	-33.5	165585	8.8
1987-88	3151	4.0	13013	14.5	693	39.7	589	32.1	181420	9.6
1988-89	3270	3.8	15229	17.0	893	28.9	786	33.9	194427	7.2
1989-90	3493	6.8	16467	8.1	852	-4.5	762	-3.4	207830	6.9
Avg. growth rate		5.11		8.18		-4.81		-4.48		8.2

$$Y_i = \beta_0 + \sum_i X_i \beta_i \quad (5.12)$$

where

X_i = number of vehicles of ith category

Y_j = consumption of fuel (kl/year)

β_i = average rate of fuel consumption per ith category of vehicles (kl/y)

The regression coefficients indicate the *marginal* increase in fuel consumption per unit increase in the number of vehicles of the ith category. If the increase is linear, then the regression coefficient is also the *average* energy consumed per month through that category of vehicles.

The resulting equations are:

For petrol-driven vehicles

$$\begin{aligned} Y = 19030 &+ \underset{(0.06123)}{0.1286}\,2W + \underset{(1.0927)}{1.2256}\,3W + \underset{(0.3166)}{0.4533}\,\text{Car} \\ &+ \underset{(0.3215)}{0.6258}\,\text{Jeep} + \underset{(1.7235)}{2.880}\,\text{Taxi} \end{aligned} \quad (5.13)$$

$R^2 = 0.98$

For diesel-driven vehicles

$$Y = 21306 + \underset{(7.024)}{15.4311}\ \text{Bus} + \underset{(2.15)}{7.5411}\ \text{Truck} \qquad (5.14)$$

$$R^2 = 0.96$$

(*Note* : The figures in parentheses are standard errors).

The results (Table 5.19) show that, on an average, a two-wheeler consumes 10.72 ltr of petrol per month whereas a car consumes 37.7 ltr. A bus consumes about 1,286 ltr per month whereas the consumption per truck is 628 ltr. The high values of *t* indicate that the regression coefficients are not zero and that there exists a relationship between the number of vehicles of different categories and the consumption of petroleum products.

Using per-vehicle-fuel-consumption-figures the total energy consumed per each category of vehicle is estimated. The results show (Table 5.20) that in the total consumption of petroleum products, the share of two-wheelers was 42.2 per cent whereas of the cars was 26.29 per cent. Trucks are the major consumers of diesel (63.31 per cent) followed by buses (27.29 per cent). In the passenger transport, buses consume 33.42 per cent of the total followed by two-wheelers with 28.22 per cent. Trucks are the major consumers of petroleum products with 40.84 per cent of the total consumption followed by buses with 17.7 per cent.

Vehicle category and energy use

In this section, energy use is disaggregated into vehicle category use. In each category, energy use is expressed as the product of a level of activity and an energy intensity. For example, the activity selected for automobiles, is vehicle-km travelled and the energy

Table 5.19: Energy consumption per vehicle.

Fuel type	Vehicle type	Regression coefficient	t value	Per-vehicle consumption (l/m)
Petrol	Two wheeler	0.1286	1.6	10.72
	Three wheeler	1.2256	1.12	102.13
	Car	0.4533	1.43	37.78
	Jeep	0.6258	1.72	52.15
	Taxi	2.8801	2.06	240.01
Diesel	Bus	15	2.05	1286.18
	Truck	8	3.2	628.43

Table 5.20: Transportation energy use by mode of transport (1989-90).

Vehicle type	Passenger transport			Freight transport	Total
	Motor spirit	Diesel	Total		
Two-wheelers	1685(42.2)		1685(28.2)		1685 (14.98)
Three-wheelers	623(15.6)		623(10.4)		623(5.54)
Cars	1050(26.3)		1050(17.6)		1050(9.33)
Jeeps	119(3.0)		119(1.7)		119(1.06)
Taxies	208(5.2)		208(3.5)		208(1.85)
Buses		1995(27)	1995(33.4)		1995(17.73)
Trucks				4595(87.5)	4595(40.84)
Others	309(7.7)		309(5.3)	668(12.5)	977(8.69)
Total	3994(100.0)	1995(100.0)	5989(100.0)	5264(100.0)	11253(100.0)

products.

$$E = VKM_i \times (EI_i)$$

where

i = Category of vehicle

E = Total fuel use (Joules)

VKM= Vehicle km

EI = Energy Intensity (Joule/km) (5.15)

Using this equation, the change in energy use over time is then decomposed into a change due to changed activity levels and a change due to changed energy intensities.

Energy consumption of any mode of transport depends mainly on its energy intensity, i.e. the inverse of efficiency (expressed as Joules per passenger km). This index is not only related to fuel efficiency but also to the use rate of modes (e.g. the loading rate of city buses) (Ross, 1989). The energy intensity of any transportation mode is calculated as:

$$EI_{ij} = \frac{F}{EF_{ij} \times LF_{ij}} \tag{5.16}$$

where

i = Category of transport

j = Type of fuel

EI = Energy intensity (J/passenger km)

EF = Fuel efficiency (vehicle km/litre of fuel)

LF = Loading factor (persons/vehicle)

F = Energy content of the fuel (J/litre)

To calculate energy intensity, one should estimate the distance travelled per vehicle category and the estimation of per vehicular consumption of fuel. Table 5.21 shows vehicle km and energy activity in detail for all modes of transportation during 1980-90. Two-wheelers accounted for the highest number of vehicle km (57.4 per cent).

Table 5.21: Vehicle type and fuel consumption (1989-90).

Vehicle type	Vehicles (no)	Fuel consumption per vehicle (kl/year)	Km/vehicle (000s)	Vehicle-kms (million)	% of total	Total fuel consumption (kl)	Energy (TJ)	% of total
Two wheeler	401811	0.13	3.86	1551	53.94	51689	1685	14.98
Three wheeler	15591	1.23	24.51	382	13.29	19110	623	5.54
Cars	71078	0.45	4.53	322	11.21	32215	1050	9.33
Jeeps	5823	0.63	3.75	22	0.76	3644	119	1.06
Taxies	2218	2.88	28.80	64	2.22	6388	208	1.85
Buses	3493	15.43	46.30	162	5.63	53911	1995	17.73
Trucks	16467	7.54	22.62	373	12.96	124180	4595	40.83
Total	516481			2875	100.00	291137	11252	100.00

Figures 5.12 and 5.13 show the shares of various transportation modes in the total energy consumption and passenger km respectively.

Between 1980 and 1990, two-wheelers achieved an annual growth rate of 25.96 per cent in passenger km and their share in energy consumption increased from 16.71 to 29.66 per cent. During the same period, the share of buses decreased from 81.89 to 76.62 per cent for PKMs, and 45.92 to 35.12 per cent for energy use (Table 5.22).

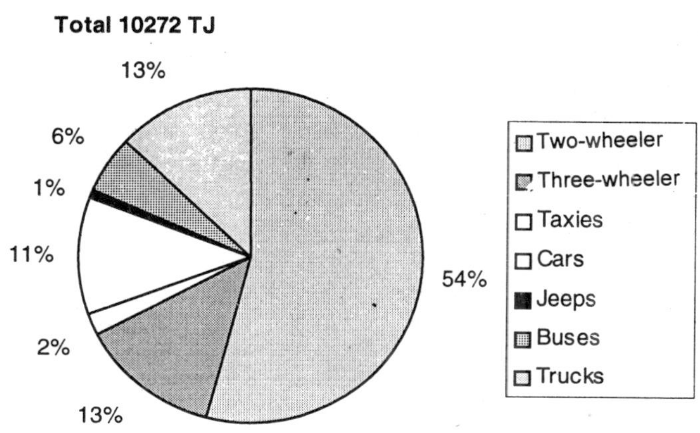

Figure 5.12: Share of transportation modes in total energy use.

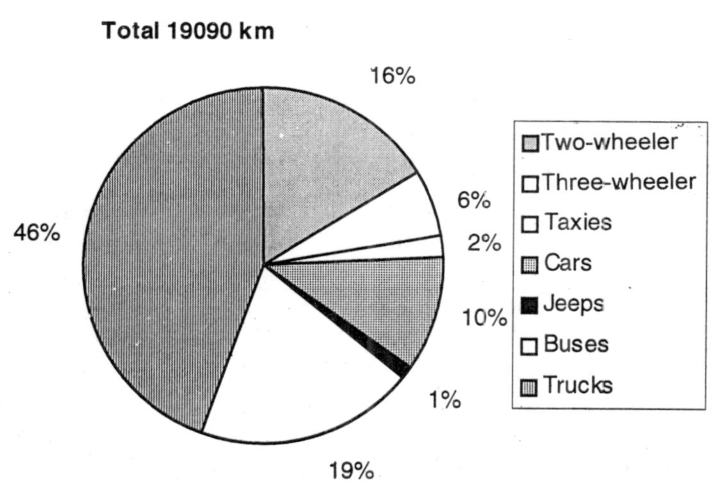

Figure 5.13: Shares of PKMs from varied transportation modes.

Table 5.22: Modes of transport and changes in passenger km and energy use.

Vehicle type	Passenger km (million)		Annual growth (%)	Share of PKMs (%)		Energy use (TJ)		Share of energy (%)	
	1980	1990		1980	1990	1980	1990	1980	1990
Two-wheelers	647	2326	25.96	6.76	12.18	469	1685	16.71	29.66
Three-wheelers	443	764	7.25	4.63	4.00	401	623	14.31	10.97
Cars	475	1063	12.38	4.97	5.56	469	1050	16.72	18.49
Jeeps	62	102	6.45	0.65	0.53	73	119	2.59	2.09
Taxies	106	211	9.91	1.11	1.10	105	208	3.75	3.67
Buses	7834	14637	8.68	81.89	76.62	1288	1995	45.92	35.12
Total	9567	19103		100	100	2805	5680	100	100

The energy intensity for different modes of transportation indicates the efficiency of any system. Table 5.23 shows the energy intensities of various modes of transport for Bangalore. An average car has an energy intensity of 0.99 MJ/passenger km and a two-wheeler 0.72 MJ/passenger km. A bus has an energy intensity of 0.16 MJ/passenger km. This wide variation between different categories of vehicles is due to various factors such as energy consumption per km, average vehicle load, speed of the vehicle, etc. Even in the same category, there will be a considerable variation depending upon the age and quality of the vehicle, traffic congestion, what work against efficiency, etc.

Table 5.23: Vehicle type and energy intensity (1989-90).

Vehicle type	Fuel consumption (kl)	Energy (TJ)	Activity unit	Vehicle-km (million)	Passenger km (million)	Energy intensity (MJ/PKM)
Two-wheelers	51689	1685	PKM	1551	2327	0.72
Three-wheelers	19108	623	PKM	344	764	0.82
Cars	32224	1051	PKM	354	1062	0.99
Jeeps	3644	119	PKM	29	102	1.17
Taxies	6388	208	PKM	70	210	0.99
Buses	65055	2407	PKM	325	14625	0.16
Total		6093		2673	19089	

Vehicle elasticity

Vehicle elasticity is an important factor in determining the impact of vehicle penetration in the future. The elasticity of a particular category of vehicle can be defined as the rate of change in total consumption of energy with respect to its percentage penetration. This can be given by the equation:

$$\eta = \frac{\Delta X_{ij} \times \beta_i}{\Delta Y_i} \tag{5.17}$$

Y_i = fuel consumption of *i*th category of vehicle (kl/year)

X_{ij} = number of vehicles in the *i*th category using *j*th fuel

β_i = average fuel consumption of *i*th category of vehicle (kl/year)

η = vehicle elasticity

This η is considered as the index representing the consumption-penetration relationship. It is dimensionless and lies between zero and one. An elasticity of zero implies that the total fuel consumption is not affected by the *i*th vehicle whereas an elasticity of one implies that the total fuel consumption is affected only by the *i*th vehicle category. The value of η in the equation indicates that a one-point increase in the penetration of a vehicle category results in a certain percentage increase in the petrol usage. For example, in the case of two-wheelers, the consumption of petrol per annum is 128.64 ltr and the vehicle penetration is 0.5023. Hence, the two-wheeler elasticity of petrol consumption is 0.47 which means that a one-per-cent increase in the penetration of two-wheelers results in a 0.47 per cent change in the consumption of petrol. Table 5.24 contains the elasticities of different vehicle categories.

Table 5.24: Vehicle elasticities.

Vehicle category	Fuel used	Per vehicle consumption (l/year)	No.of vehicles	Vehicle penetration	Elasticity
Two-wheelers	Petrol	128.64	401811	0.5023	0.470
Three-wheelers	"	1225.68	15591	0.0195	0.174
Cars	"	453.24	71078	0.0888	0.293
Jeeps	"	625.80	5823	0.0073	0.033
Taxies	"	2880.12	2218	0.0028	0.058
Buses	Diesel	15433.44	3493	0.0044	0.259
Truck	"	7541.14	16467	0.0206	0.598

On this basis, a one-per-cent increase in the penetration of two-wheelers can be estimated to lead to an increase in Bangalore's petrol consumption to 387.90 kl per year. Similar calculations have been carried out to estimate the effects on petrol and diesel consumption of any increase in the penetration of other vehicles (Table 5.25)

Table 5.25: Effect of increase in vehicle penetration on fuel consumption.

Vehicle category	Vehicle penetration	Elasticity	Increase in consumption (kl/m)
Two-wheelers	0.5023	0.470	387.90
Three-wheelers	0.0195	0.174	190.26
Cars	0.0888	0.293	320.75
Jeeps	0.0073	0.033	36.28
Taxies	0.0028	0.058	63.60
Buses	0.0044	0.259	537.45
Trucks	0.0206	0.598	1238.02

Interdependence

The interdependence of one vehicle on another is an important factor that must be given due consideration. A ratio of the number of one type of vehicle to another can be formulated as given in Table 5.26. As the table shows, the ratio of the number of cars to the number of two-wheelers decreased even though there was an individual increase. In

Table 5.26: Change in the passenger transport vehicles.

Year	No.of vehicles					Ratio of vehicles						
	Two-wheelers	Three-wheelers	Cars	Taxies	Buses	3W/ 2W	Car/ 2W	Taxi/ 2W	Bus/ 2W	3W/ Car	Bus/ 3W	Bus/ Car
1980-81	111750	10046	31738	1120	4671	0.090	0.284	0.010	0.042	0.317	0.465	0.147
1981-82	125600	10355	32429	1171	5090	0.082	0.258	0.009	0.041	0.319	0.492	0.157
1982-83	135536	11063	33506	1276	5507	0.082	0.247	0.009	0.041	0.330	0.498	0.164
1983-84	162206	11880	36064	1607	5931	0.073	0.222	0.010	0.037	0.329	0.499	0.164
1984-85	189619	11348	47787	2027	6155	0.060	0.252	0.011	0.032	0.237	0.542	0.129
1985-86	216234	10446	53150	1756	5168	0.048	0.246	0.008	0.024	0.197	0.495	0.097
1986-87	236726	10524	54885	1717	5704	0.044	0.232	0.007	0.024	0.192	0.542	0.104
1987-88	301677	12439	64703	2027	6575	0.041	0.214	0.007	0.002	0.192	0.529	0.102
1988-89	343773	15513	65916	2089	4247	0.045	0.192	0.006	0.012	0.235	0.274	0.064
1989-90	401811	15591	71078	2218	4215	0.039	0.177	0.006	0.010	0.219	0.270	0.059

the case of the number of cars to autos, and two-wheelers to autos, the ratio increased. This indicates that two-wheelers are increasing faster as compared to other categories of vehicles. In the case of the public transport service, the ratio is decreasing with all other categories of vehicles indicating that it is not keeping pace with the other means of transport.

Discussion

Transportation and energy demand

Energy demand in transportation varies with the following factors: income, population, passenger km, and the energy intensity of various transporation modes (Sathaye *et al.*, 1987). Based on the above factors, the structure of energy demand for road transportation can be studied using the following equation:

$$\underset{\text{(travel demand)}}{PTE_t = PKM_{ijt}} \times [\underset{\text{(Capital stock)}}{(V_{ijt})} \times \underset{\text{(Intensity of use)}}{(KM/V)_{ijt})/PKM_{ijt}]} \times \underset{\text{(Energy intensity)}}{[(E/VKM)_{ijt})]} \qquad (5.18)$$

where

i = Index for the vehicle category

j = Index for the fuel type

t = Index for the year

PTE= Passenger Transport Energy Demand (GJ/year)

P = Passengers traveled (no)

KM/P= Intensity of transport demand (km/passenger)

PKM= Passenger kilometres demanded = $P * (KM/P)$ (km/yea

V = Number of vehicles in operation

KM/V= Intensity of vehicle use (km/vehicle)

VKM = Annual vehicle kilometres

E = Energy consumed (GJ)

E/PKM = Energy intensity factor (GJ/PKM)

Using equation (5.18), the energy demand, PTE_t, can be disaggregated into different factors.

The first factor, V, is the number of vehicles. With increasing incomes and availability of vehicles on loan and lease, the rate of growth of vehicle population increased rapidly over the years. The average number of two-wheelers per family was 0.185 during 1981 which increased to 0.45 in 1990 (Reddy, 1990). This indicates that one out of every two households in Bangalore possessed a two-wheeler. The respective figures for cars were 0.05 and 0.08. This is due to the fact that, with increasing urbanisation, the need and demand for passenger transport increases simultaneously. Also, with an increase in population, the number of people travelling increases, thereby increasing the number of vehicles. Thus, with increasing urbanisation, and population, not only the P_{ijt} increases with respect to t, but the proportion of P_{ijt} allocated among different categories of j changes with respect to the growth rate of population, availability of the vehicles, income of the population, rate of urbanisation, and the relative transport fuel prices. The second factor, km/V, is the average run per vehicle per year. These values are 4,530 km/year for cars and 3,860 km/year for two-wheelers. With the almost explosive growth of urban centres, the distance between the home and work place increases, thereby increasing the distance travelled.

The last factor to be considered, E, is the fuel consumption. This depends on the energy efficiency of the mode of transport which can be measured by the energy input required to transport passengers over a given unit of distance. Thus, the efficiency of a vehicle can be defined in terms of $(E/VKM)_{ijt}$, or the product value of load factor $(VKM/PKM)_{ijt}$ and the energy intensity of a vehicle can be defined through $(E/PKM)_{ijt}$. For a stated vehicle energy intensity and energy efficiency can be maximised if the vehicle is operated at 100 per cent passenger capacity. Thus when a private car is driven by one person or when the public transport buses run without the full capacity, the energy intensity per passenger km increases. At present norms, the fuel energy intensity of a car is 0.99 MJ/PKM whereas for a two- wheeler, it is 0.72 MJ/PKM and for a bus 0.16 MJ/PKM. With the introduction of many fuel efficient two-wheelers and cars, the consumption per vehicle seems to have come down. However, this fact is yet to be tested.

As the above results show, in the near future, two factors will play an important role in the reduction of energy demand and one factor will contribute to its acceleration. The former is improvement in the efficiency of the vehicle engine and less average run per vehicle and the latter is an increase in the total number of vehicles.

Impact of transportation on environment

Environment quality concerns are increasing due to the fact that (a) the quality of metropolitan air continues to be unsatisfactory in many areas, and the public is clearly interested in making progress, (b) regional air quality impacts, especially acidification of lakes and denudation of forest are creating increasing problems and (c) the greenhouse effect which has affected the global climate, as a result of the increasing atmospheric concentration of infra-red radiation absorbing gases, such as CO_2, NO_x, methane and chlorofluoro carbons.

In Bangalore, transportation energy, particularly by road vehicles, constitutes nearly 90 per cent of the total energy consumption out of which 80 per cent is through passenger transport. Between 1980–90, there was a significant growth in personal transport, particularly by two-wheelers (15.41 per cent p.a.) which accounted for 60 per cent of the total vehicle kilometres. In energy use the share of two-wheelers and cars was around 25 per cent of the total. This growth in personal transport, which has a high energy intensity, will increase energy consumption. As our results suggest, the energy intensities of each mode of travel could have a significant impact on energy use and thereby on the environment.

Carbon monoxide, un-burnt hydrocarbons, etc., emitted from vehicles contribute to the atmospheric pollution. The carbon monoxide resulting from consumption of petroleum products is inversely proportional to their fuel economy. Exhaust emissions from a petrol-driven engine consist mainly of carbon monoxide, un-burnt hydrocarbons, oxides of nitrogen and partial oxidation products of the aldehyde family. In diesel-driven engines, the concentration of CO and un-burnt hydrocarbons in the exhaust is low, while nitrogen oxides are present in high concentrations (Pundir *et al.,* 1985).

Table 5.27 presents the vehicle emission inventory for Bangalore. An important observation from these data is that the contribution of unburnt hydrocarbons by two-wheelers is very high (50 per cent of the total). In the case of diesel-driven vehicles, CO constitutes 13 per cent of the total.

Figure 5.14 shows the contribution of emissions for different categories of vehicles and Figure 5.15 gives detailed information on the relative impact of different modes of transport on energy consumption and environment.

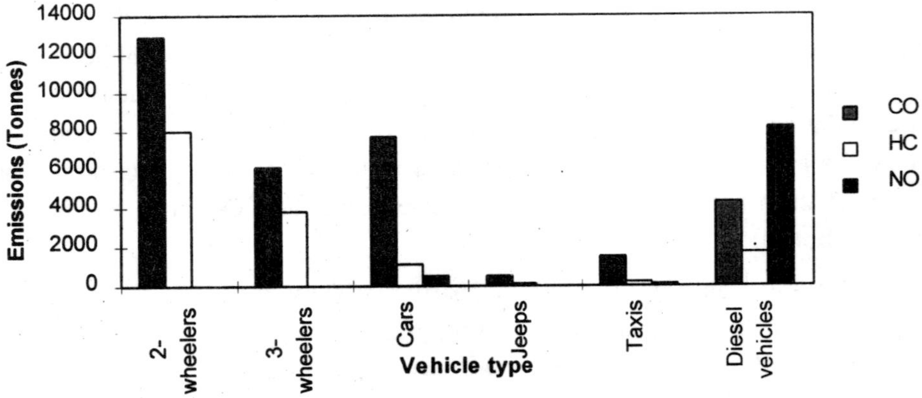

Figure 5.14: Emissions from various modes of transport.

Table 5.27: Vehicle emission inventory for Bangalore (1989-90).

Type of vehicle	Average mass emissions (g/km)			Vehicle-km (million)	Total emissions (tonnes/year)		
	CO	HC	NO_X		CO	HC	NO_X
2-wheelers	8	5	-	1551	12873	8034	-
3-wheelers	16	10	-	382	6051	3770	-
Cars	24	4	2	322	7738	1150	506
Jeeps	24	4	2	22	529	79	35
Taxies	24	4	2	64	1538	228	100
Diesel vehicles (*)	0	0	0	189234	4315	1665	8156
Total					33043	14926	8797

* g/g of fuel consumed.

Since economic growth, technological advancement and transportation are inter-related, the increase in transportation's share of energy will be quite high in the coming years. As the vehicle elasticity results indicate, a one percentage point increase in the two-wheeler population results in a 0.47 per cent increase in petrol consumption. Thus, with the increasing use of transport services, the level of energy consumption, and thereby the atmospheric pollution rises. This poses a major environmental dilemma.

When compared with vehicles in other countries, Indian vehicles are high emitters of pollutants like CO, HC, and NOx. Any programme for the control of vehicle

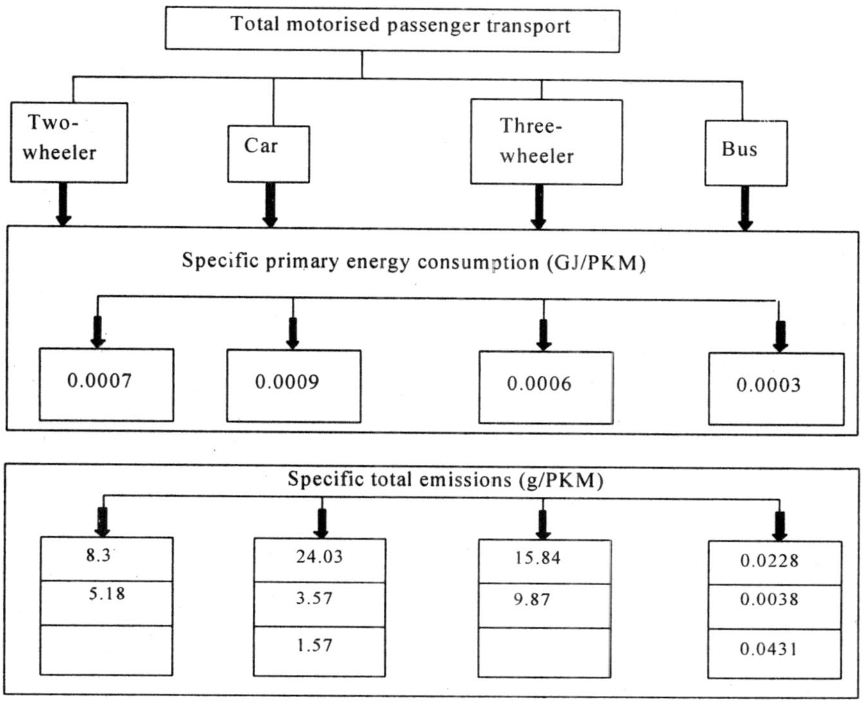

Figure 5.15: Impact of different modes of transport on environment.

emissions must be directed at the reduction of emissions, mainly from two-wheelers as they constitute the single largest group in the vehicle population.

A major fuel economy tied to emission reduction arises from the nature of the mass regulatory standards for emissions, i.e. the limits on grams of emission per km travel. It is easy for a vehicle that consumes relatively little fuel per km to meet the standard in percentage terms (concentration of pollutants in the combustion gases). Emissions decrease less than a simple proportion of fuel use would suggest. For example, the use of a smaller engine.

In addition, some fundamental approaches to fuel economy improvement also enable percentage emission reductions, e.g. in an efficient engine, the combustion temperature is lower reducing NO_X formation. The fuel efficiency is nevertheless improved, because air is a better thermodynamic medium than (evaporated) petrol. The increase in unburnt hydrocarbons, with air to fuel ratio, can be prevented, in principle, by improving ignition and can be mitigated by improved exhaust after-treatment. In the long run, vehicle-emission standards should be formulated on the basis of pollutants emitted by the different categories of vehicles under actual driving conditions. Necessary test procedures should be developed and the requisite facilities established. Assessment of the present engine technology, monitoring of air quality and establishment of laboratories for vehicle certification for emissions should be made obligatory.

Actually, automobiles cause pollution in several ways and control of engine emissions deals with but one of these. For example, an auto tyre reduces its weight by about 1/3 kg every year, the loss being in the form of rubber dust abraded away against the road. Thus, each road vehicle contributes some 1/3 kg of rubber dust to the ambience each year. While there has been no definite study of the flow and ultimate end of this pollutant, it can be assumed that a portion of it goes into the air we breathe while some is washed down to earth, perhaps to pollute the drinking water (Whitelegg, 1991).

Asbestos dust is a known, identified carcinogen. Every time an automobile's brake pedal is pressed, asbestos dust is released into the atmosphere. Autobrake linings are made up of asbestos, and in the course of converting the machine's kinetic energy into heat, these linings are converted into asbestos dust.

Any programme for the control of vehicle emission must be directed towards the reduction of emissions from two-wheelers which are generally powered by two-stroke engines due to their low cost. On the other hand, four-stroke engines provide higher thermal efficiency and emit lower quantities of hydrocarbons (1/5th to 1/10th of the emissions from two-stroke engines). Therefore, if all vehicles are powered by four-stroke engines an approximately 50 per cent reduction in hydrocarbon emissions can be achieved. Further, mandatory emission inspection and maintenance programmes have to be introduced to improve the air quality in urban areas.

If there is an increase in the petroleum prices, the level of travel may decrease, thereby decreasing the energy intensities of each transportation mode. However, with increasing incomes, personal transport will increase as also the per capita energy consumption. Thus, the impact of the growing volume of personal vehicles more than offsets the impact of reduced energy intensities. Modal shifts to less-energy-intensive modes (from two-wheelers to bus) could contribute to energy savings. To achieve this, the government should make the public transport bus system more efficient and seriously consider the rail transit system. Pricing of transport services based on some sort of bus operating index, incentives for economic operations, fuel conservation, punctuality, etc., will improve the performance of buses.The introduction of point-to-point travel by public transport, education on conservation of fuel, etc., are also important factors to be considered. Fuel efficient vehicles should be encouraged with some incentives in the form of tax concession, etc. Most importantly, the environmental consequences of transportation should be given importance while framing transportation policies.

Demand side management

Technological

Since 80 per cent of the transport energy is through two-wheelers and cars, a shift to small cars and lighter two-wheelers, which are fuel economic, could save significant quantities of fuel. The major fuel economy improvements in the past decade in this section are: (i) propulsion system efficiency_engineering improvements give a 36 per cent increase in power/unit of engine size; (ii) vehicle design_improved design and use of new materials reduce the ratio of weight to interior volume of vehicles, thereby reduce the fuel consumption; (iii) trade-offs_trade-offs among the performance, emissions, cost, safety and fuel economy have been used by the manufacturers in meeting this.

In Bangalore, 90 per cent of the total transport energy is through road vehicles. Thus, there is a clear need to reduce the energy for road transport which depends on oil-based fuels and this requires changes in engine and motor technology. The most promising alternative technologies appear to be Compressed Natural Gas (CNG) and electric motors. CNG is obtained by compressing gas to 2300-2600 psi. This is most suited for vehicles with high mileages and restricted range of operations such as buses and taxis, and thus the urban areas are most suited for CNG operations (Moreno and Bailey, 1989; Painuly and Parikh, 1994). Electric vehicles require a wire feed or heavy batteries to provide power. Since wayside connections are not feasible for most vehicles, electric vehicles can be battery-powered. Improved battery vehicles will result in traffic compatible vehicles capable of limited application such as in urban areas (Scott, 1982).

The improvement in electronic communications has reduced the travel needs of people, thereby reducing the consumption of petrol and diesel. Many financial transactions requiring transport of scores of sheets of paper can be done much faster using the improved telecommunication systems. Remote work stations, armchair shopping, train and air reservations, tele-conferences, etc., allow the user to avoid physical travel.

To develop and exploit technological opportunities to further reduce fuel consumption, it would be valuable to strengthen technological policies (such as R&D programmes) as well as enact still more effective air pollution standards. In designing more effective standards, greater attention should be paid to (a) fuel economy - emission interaction and (b) non-standard situations. Also taxes and rebates based on emission performance might be effective.

Policy formulations

(a) Planning: Transport planning policies offer greater scope for reducing energy consumption. Since two-wheelers and cars consume 50 per cent of the total energy for passenger transport, shifting of modes of transport from two-wheelers and cars to buses could save energy. If one passenger km. is shifted from a high energy intensity mode to a low one, the savings will vary from 0.49 to 1.32 MJ (Table 5.28).

Also, the long-awaited rail transit system is urgently needed since the present bus system cannot sustain the increasing load. Land planning is also an important factor to be considered

(b) Pricing: Generally, fuel price increases motivate fuel economy, but, as the recent trends suggest, the effect is not thought to be strong. Therefore, alternative pricing strategies should be given importance. This should go further than just road pricing. There should be ecological taxation reform which shifts the present taxation burden progressively over a 30-year period away from income tax. This should penalise those activities which consume space, energy and create pollution leading to ecological and environmental degradation. If applied, these policies can steer the economy towards a state which encourages environmentally friendly activities and discourages the opposite.

Table 5.28: Reduction in energy use from shifting one mode of transport to another.

Change in transportation mode	Energy intensity (MJ/PKM)		Saving (MJ)
	One mode	Another mode	
Two-wheeler to bus	0.72	0.16	0.56
Three-wheeler to bus	0.82	0.16	0.66
Car to bus	0.99	0.16	0.83
Jeep to bus	1.17	0.16	1.01
Taxi to bus	0.91	0.16	0.75
Car to three-wheeler	1.09	0.82	0.27

An effective strategy for reducing the consumption of petroleum products must focus, therefore, on the transportation sector and on vehicular consumption of these products. Such a strategy would include an emphasis on mass transportation and more efficient vehicles.

A developing country like India cannot afford to neglect the development of an efficient transportation system.

6

ELECTRICITY

I 'saw' cascades of energy coming down from outer space,
in which particles were created and destroyed in rhythmic pulses;
I 'saw' the atoms of the elements and those of my body
participating in the cosmic dance of energy; I felt its rhytm and
I 'heard' its sound, and at that moment I knew that
this was the dance of Shiva, the lord of Dancers.

– Pritjof Capra in the Tao of Physics

India's power planning has always had a strong supply bias. Traditionally, utility (electricity board) planners forecast the load and compare it with the existing generation capacity. Based on this comparison, and relatively well-established rules and techniques, the planners produce a long-range plan to meet the needs of customers. However, it was felt that there was a strong need to review this planning process. To obtain a realistic picture of the electricity demand one has to obtain the demand for different end-use services. This is because, the end-use services vary with the stock of electrical appliances and the efficiencies with which they are being used.

In this chapter the consumption of electricity in various sectors is studied. For the residential sector, a theoretical framework is built within which the demand for electricity is studied. The cross-sectional household survey data are analysed to estimate the consumption of electricity for different end-uses by relating the household electricity consumption with the stock of electrical appliances. Using these estimates, appliance elasticities are obtained. It is shown here how these elasticities can be used to relate the future electricity usage to changes in the stock of appliances for the city of Bangalore.

It is common practice to use price elasticities for predicting future energy demands (Taylor, 1975). However, it is felt that price elasticities, which are difficult to estimate, do not present a correct picture because of the hidden subsidy in the energy prices. Even in countries where a great deal of research on electricity demand is being carried out, nothing much is known about price elasticities and the time lags in demand responses. To obtain a realistic picture of the electricity demand for different end-use services, the composition of appliance stock plays an important role (Dubin and McFaden, 1984). This composition can be influenced by household income, household size, etc.

There have been a number of econometric studies of residential electricity consumption. Wilson (1971) and Anderson (1973) used cross-sectional data to estimate the effect of household characteristics and price of electricity on the household consumption and concluded that price is the major determinant of electricity consumption. Houthekar's (1973) results yielded the short-run income and price elasticities, i.e. the effect of income and price on consumption with the stock of appliances being held fixed. The best known modern empirical study was published by Fisher and Kayson (1962). In their analysis of residential electricity consumption, they investigated demand variations, price and income elasticities and concluded that the residential electricity demand is proportional to the stock of appliances. Parti and Parti (1980) disaggregated the total household consumption into appliance-wise consumption using regression analysis. They also estimated the effect of exogeneous variables such as family income, residence size, etc., on appliance-wise consumption.

It has been observed that increasing urbanisation, increase in monetary incomes of households and availability of modern energy carriers like LPG and electricity can be attributed to the carrier shifts (Reddy, 1990). In the case of electricity, this has resulted in the growth of electrical appliances. Two sets of questions can be posed. What is the role of appliance stock in electricity consumption? How much energy-saving could come about by changes in the appliance stock and through altering the technological nature of appliances designed to conserve energy?

Electricity conservation is a topic that is discussed in India, in qualitative terms or in terms of percentage savings possible by different options. In order to give conservation the role it deserves in strategic planning, quantification of targets in terms of costs and energy and power savings, as well as the distribution of benefits and costs among stakeholders, viz. consumers, utility and society, is needed. In the Eighth Plan (1993-97), the need for financial support for energy conservation and management was recognised and funds were allotted in general terms (Anon, 1992). Yet, a methodology to allocate these funds is not so far available.

The cost of electricity conservation is less than that of generation. This is because, in the short run, conservation programmes have short gestation periods, save fuel, improve the quality of service and are environment friendly. In the long run, there are savings in capacity additions which are otherwise highly investment intensive. Through conservation, the customers benefit from reduced energy bills while utilities would save on fuel and capacity. The savings for society will be in terms of availability of more funds for social welfare, and less pollution (Hirst and Meiers, 1973; Geller, 1991; Busch, 1992). Of course, conservation has a limited but important role and cannot substitute for new capacity for a developing country but only reduce the need for expensive new capacity.

This chapter contributes to the analysis of electricity demand in many ways particularly in the residential sector. Firstly, this is the only study of its kind attempting to analyse the micro-level data using an appliance census approach to estimate the appliance consumption which is used to relate the electricity use to changes in appliance penetration levels. The appliance elasticities that are estimated here play an important role in predicting the future electricity consumption. Secondly, it also provides data on specific conservation options and energy and power savings through these options and their environmental benefits.

Electricity Supply, Transmission and Distribution

Bangalore gets its electricity mainly from the Shravati Hydroelectric Power Station which is at a distance of about 350 km from Bangalore; supplies are also obtained from other power stations in Karnataka and from the neighbouring stations of Andhra Pradesh. Most of Karnataka's power is generated by the Karnataka Power Corporation (KPC) Ltd (Anon, 1992). The generated electricity flows through grid transmission lines to the switching stations and from there to the transformers for distribution to the consumers.

The transmission, distribution and sale of electricity is controlled by the Karnataka State Electricity Board (KEB). Thus, KEB purchases power from KPC and sells it to the consumers.

Consumption

The consumers of electricity are broadly classified into the following six categories: (i) residential, (ii) commercial establishments, (ii) industrial consumers – low-tension (LT), (iii) industrial consumers – high-tension (HT), (iv) street lighting, (v) irrigation and water supply and (vi) temporary power supply to functions, meetings, etc.

Bangalore's demand for electricity has been increasing over the years. The consumption in 1980-81 was 1,027 Giga Watt hours (GWh) which increased by a factor of 2 to 2,290 GWh during 1989-90 (Table 6.1).

To understand the annual consumption (*E*) as a function of time, the total electricity consumption was regressed on time in years (*T*) which indicates that the consumption is fairly elastic at an average annual growth rate of 9.13 per cent and the *t* statistic given in parentheses is significant at 1 per cent level.

$$\ln (E) = 6.86 + \underset{(15.77)}{0.0913\, T} \qquad (6.1)$$

$R^2 = 0.98 \quad DW = 1.45$

(The figure in parentheses is the *t* value)

There is also a change in the consumption pattern of different categories (Figure 6.1). During the period 1980–90, the residential electricity consumption (RES) increased at a rate of 16.63 per cent annually, commercial consumption (COM) at an annual rate of 7.4 per cent, and industrial consumption (IND) at 4.1 per cent and others

Table 6.1: Electricity consumption (GWh) by the end-use sector (1980-90).

Year	Residential	Commercial	HT	LT	Street lighting	Temp. power	Water supply	Total
1980-81	306.6(29.8)	42.5(4.1)	634.3(61.8)	11.8(1.2)	21.3(2.1)	1.9(0.1)	3.8(0.4)	1027(100.0)
1981-82	365.2(30.4)	45.0(3.8)	745.9(62.4)	13.7(1.2)	22.8(1.9)	2.4(0.2)	4.0(0.3)	1199(100.0)
1982-83	429.1(34.1)	46.8(3.8)	729.6(58.5)	15.6(1.3)	23.8(1.9)	1.7(0.1)	4.4(0.4)	1257(100.0)
1983-84	529.7(39.3)	47.4(3.6)	711.5(53.5)	16.5(1.3)	24.4(1.9)	1.7(0.1)	4.6(0.4)	1348(100.0)
1984-85	613.8(43.1)	51.5(3.6)	719.2(50.6)	17.5(1.2)	24.7(1.7)	2.5(0.2)	4.9(0.4)	1423(100.0)
1985-86	774.1(45.1)	66.5(3.9)	810.9(47.2)	20.5(1.3)	31.2(1.8)	3.2(0.2)	6.9(0.4)	1715(100.0)
1986-87	884.3(47.5)	68.2(3.6)	845.1(45.4)	19.3(1.1)	33.0(1.8)	4.2(0.2)	6.2(0.3)	1860(100.0)
1987-88	925.4(47.4)	66.1(3.4)	897.5(46.0)	21.3(1.1)	32.4(1.6)	4.2(0.2)	6.3(0.3)	1953(100.0)
1988-89	993.2(47.1)	74.5(3.5)	965.1(45.8)	25.4(1.2)	36.7(1.7)	4.9(0.2)	6.7(0.3)	2108(100.0)
1989-90	1076 (47.9)	83.5(3.6)	1024.2(44.7)	26.7(1.2)	39.1(1.7)	5.5(0.2)	7.2(0.3)	2290(100.0)

Note: Numbers in parentheses represent percentages.

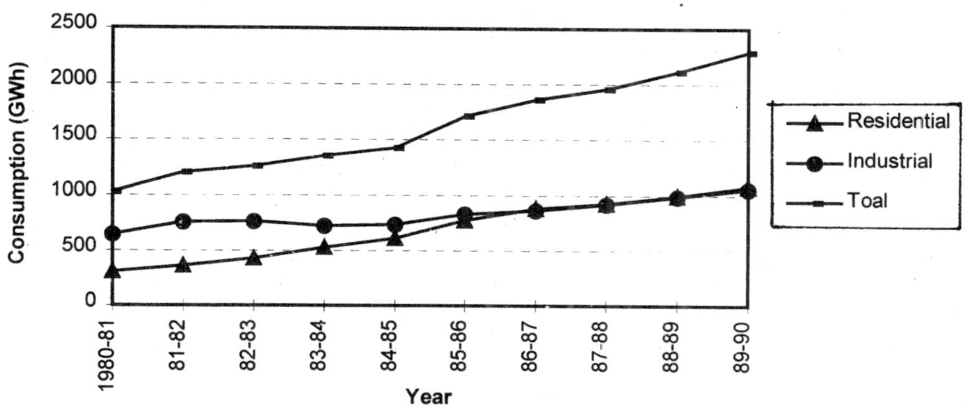

Figure 6.1: Pattern of electricity consumption (1980-90).

(OTH) at 7.24 per cent. While the share of the residential category increased from 30 to 47 per cent, the industrial category's share fell from 63 to 47 per cent. No change was noticed in the consumption pattern of the commercial category which stood around 4 per cent in the same period (Figure 6.1). The growth rates are based on the equations presented below.

$$\ln (RES) = 5.5866 + 0.1663\ T \quad (6.2)$$
$$(21.86)$$
$$R^2 = 0.99 \quad DW = 1.35$$

$$\ln (COM) = 3.64 + 0.0741\ T \quad (6.3)$$
$$(6.8)$$
$$R^2 = 0.88 \quad DW = 1.3$$

$$\ln (IND) = 6.47 + 0.041\ T \quad (6.4)$$
$$(5.08)$$
$$R^2 = 0.81 \quad DW = 1.53$$

$$\ln (OTH) = 3.20 + 0.0724\ T \quad (6.5)$$
$$(7.08)$$
$$R^2 = 0.89 \quad DW = 1.5$$

(Figures in parentheses are *t* statistics)

The regression coefficients are significant at a 1 per cent level. The values of the Durbin-*Watson statistic* indicate that the null hypotheses of no serial correlation cannot be rejected at a 5 per cent level. As the data show, residential consumption accounted for about 46 per cent of the total electricity consumption and is growing at a rate of 17 per cent per annum. Therefore, it is important to study the residential electricity consumption in a greater detail.

Residential Consumption

Residential consumption is categorised into (i) non-all electric homes (non-AEH) (single- phase supply), (ii) all electric homes (AEH) (three-phase supply; current supply > 15 amperes), (iii) KEB employees who get a 'free' supply within limits. Electrical energy consumption in the residential sector in Bangalore is increasing at a rapid rate. It increased from 310 to 1,076 GWh from 1980 to 1990. The average monthly consumption was 26 GWh during 1980 which increased to 91 GWh during 1990, a three-fold increase. In the residential sector, electricity is supplied to non-AEH and AEH categories. Between 1980 and 1990, the non-AEH consumption more than doubled from 58.3 to 135 GWh, while the AEH consumption increased four-fold from 248 to 1,040 GWh. The following equations fit the consumption (GWh) *vs* the years data.

$$\ln(\text{non-}AEH) = 3.91 + \underset{(9.98)}{0.1073}\,T \qquad (6.6)$$

$R^2 = 0.94 \quad DW = 1.57$

$$\ln(AEH) = 5.37 + \underset{(6.68)}{0.1782}\,T \qquad (6.7)$$

$R^2 = 0.98 \quad DW = 1.3$

(Figures in parentheses are *t* statistics)

The equations show that the non-AEH consumption increased at an average annual rate of 10.73 while the growth rate of AEH was 17.82. The values of R^2 are very high and the *t* statistics are significant at a 1 per cent level. The value of the Durbin-Watson statistic indicates that there is no serial correlation. Figure 6.2 shows the trend in the consumption of electricity by AEH and non-AEH households.

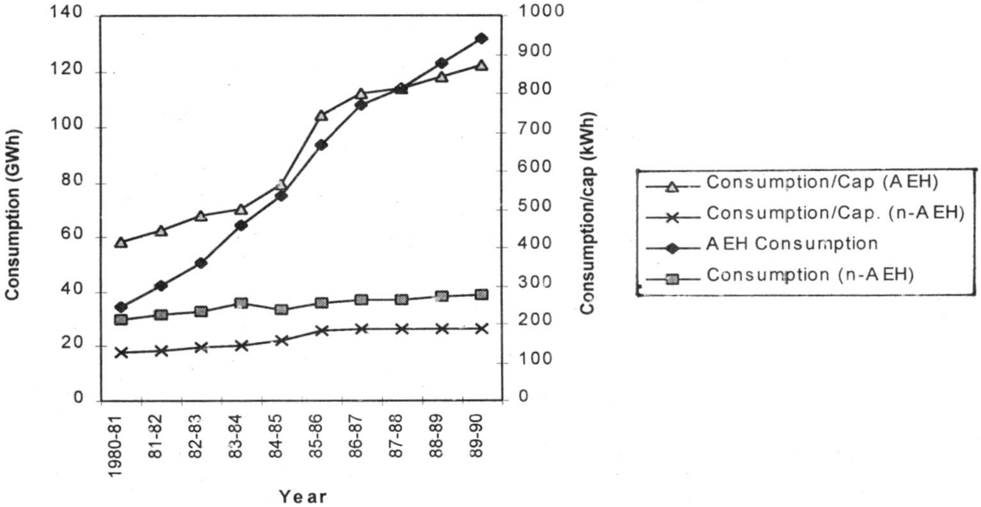

Figure 6.2: Pattern of residential electricity consumption (1980-90).

Among the electrified houses, the increase in the case of non-AEH houses, from 0.27 million in 1980 to 0.43 million in 1990, stood at 35 per cent while that of AEH households stood at 300 per cent, increasing from 0.097 million in 1980 to 0.31 million in 1990. The consumption per household also increased from 214.3 to 249.8 kWh per month in AEH and from 18.3 to 27.4 kWh per month in non-AEH from 1980 to 1990.

AEH and non-AEH household characteristics

Out of the 1000 households covered during the survey, 310 are AEH, 500 are non-AEH and 190 households do not have any electricity connection. Though the average family size is only 7.5 per cent more in non-AEH (5.16) households than in AEH (4.80) ones, the differences in income between the two groups are very large. The average monthly income of an AEH household is roughly three times that of a non-AEH household. Further, the dwellings of AEH homes are also much larger, having roughly twice as much area as the non-AEH dwellings. In the case of monthly electricity consumption, the average consumption of 215 kWh per month in the AEH home is eight times that of a non-AEH home (Table 6.2).

Table 6.2: Household characteristics in the sample.

Variable	Non-AEH		AEH		All households	
	Average	Std.dvn	Average	Std.dvn	Average	Std.dvn
Family size (no.)	5.16	2.90	4.79	2.26	4.95	2.56
Family income (Rs/m)	1476.24	1158.12	4127.19	4643.95	2987.50	3819.85
Rooms(no.)	2.73	1.19	4.82	2.26	3.92	2.15
Area (sq.ft)	293.49	170.28	640.22	487.33	491.13	421.07
Electricity consumption (kWh/m)	29.53	15.18	214.89	86.97	136.91	111.87

There is an enormous difference in the number of electrical appliances of various types in each category owned by AEH and non-AEH households (Table 6.3). The survey data indicate that appliances such as electric stoves, water-heaters, refrigerators and pumps (to lift water from sump to an overhead tank) are special features of AEH households whereas incandescent bulbs, fans, television sets and radios are common to both the groups. Water-heaters figure in almost all the AEH households with one or two exceptions. Seventy five per cent of the AEH households own an electric stove and 70 per cent television sets. Even though bulbs are a common feature in both AEH and non-AEH, their number per household varies from 8.9 bulbs per AEH house to 4.4 bulbs per non-AEH house. In the case of tubelights, the average number per household is 0.61 in the non-AEH whereas in the AEH it is 2.1. The same is true for other appliances such as ceiling fans, mixies (blenders/grinders), etc. Variations in the appliance ownership are assumed to be due to the differences in income. Therefore, an appliance stock can be considered as a proxy for income.

Approaches to the study of residential electricity demand

Even though the total electrical energy consumed in the residential sector is known, the quantity used for different applications like lighting, cooking, water-heating, etc., has not been determined thus far. Unlike firewood, kerosene and LPG, electricity can be used for a variety of applications or end-uses ranging from cooking, water-heating and lighting to air circulation and refrigeration. It is important to determine the consumption of electricity by these different end uses.

Table 6.3: Appliances stock in the sample.

Appliance category		Appliances per household	
		Non-AEH	AEH
25 Watt Bulb	(25 B)	0.36	1.03
40 Watt Bulb	(40B)	2.47	3.78
60 Watt Bulb	(60B)	1.35	2.91
100 Watt Bulb	(100B)	0.43	1.18
Tubelight	(TB)	0.61	2.10
Elec. Iron	(EI)	0.43	0.89
Ceiling Fan	(CF)	0.51	1.87
Table Fan	(TF)	0.32	0.54
Grinder	(GR)		0.17
Mixie	(MX)		0.51
Motor	(MT)		0.13
Radio	(RD)	0.63	0.84
Refrigerator	(RF)		0.24
Stove	(ST)		0.75
Television	(TV)	0.36	0.70
Waterheater	(WH)		1.01

The approaches to the study of residential electricity demand can be categorised as follows: (i) a conventional econometric approach, (ii) an engineering approach and (iii) an appliance stock approach. According to the conventional approach, the quantity of electricity consumed is studied as a function of economic variables such as family income, size of the family, residence size, price of electricity and its substitute. The engineering approach is based on the results of sample surveys of variables which permit a calculation of the electricity used for different applications. The appliance stock approach is based on the assumption that the stock of electrical appliances of different categories is the major determinant of electricity consumption.

(i) Conventional approach

Household demand behaviour may be expected to reflect not only income but also other household characteristics and the environment in which they operate. The basic question is whether there is a significant relationship between the electricity consumed and other econometric variables describing the household.

Usually, electricity consumption is considered as a function of family income, family size, residential area, price of electricity and its substitute. Then, the form of the equation is

$$E_i = \Sigma_k V_{ik} \times b_k \tag{6.8}$$

where V_{ik} represents the value of $_k$ exogeneous variables of the ith household and b is a constant.

The determinants of electricity consumption considered in this study are family income (FI), family size (FS) and residence size (AREA). The price of electricity was not included in the model, because a cross-sectional data base does not yield information on the effects of price variations. When the logarithmic values of per household consumption were regressed on the price of electricity for a period of five years, the price elasticity of demand (0.45) shows that the consumption per household increases irrespective of the price variation.

When the electricity consumption (E) was regressed on family income (FI) (Rs/month), family size (FS) (number) and residence size ($AREA$) (sq.ft), the following results were obtained.

$$E = 62.074 + \underset{(1.509)}{3.744\,FS} + \underset{(0.002)}{0.005\,FI} + \underset{(0.016)}{0.159\,AREA} \tag{6.9}$$

$R^2 = 0.54$

$$E = 79.81 + \underset{(0.001)}{0.0192\,FI} \tag{6.10}$$

$R^2 = 0.38$

(Figures in parentheses represent standard errors.)

The value of R^2 indicates that only 54 per cent of the variation in electricity consumption is explained by the variables FS, FI and $AREA$. And if FI is considered separately, it explains only 38 per cent of the variation.

As expected, all the coefficients are positive. This is because the electricity consumption increases with an increase in FI, FS and AREA. The partial coefficients of determination and the regression coefficients indicate that these variables are not the dominant determinant of electricity consumption.

(ii) Engineering approach:

The total electricity consumption in a household $_i$ is related thus to the appliance-wise consumption:

$$E_i = \Sigma_j \; E_{ij} \tag{6.11}$$

where E_{ij} is the electricity consumed by the $_j$th appliance category of the ith household. Also,

$$E_i = \Sigma_j \; U_{ij} \times X_{ij} \tag{6.12}$$

where U_{ij} is the electricity consumption of the $_j$th appliance category (kWh/month) and X_{ij} is the number of electrical appliances in the $_j$th category. U_{ij} can also be written as

$$U_{ij} = w_{ij} \times h_{ij} \tag{6.13}$$

where w_{ij} is the wattage of the $_j$th appliance category and h_{ij} is the number of hours per month for which the $_j$th appliance is being used. This U_{ij}, referred to as the engineering estimate of the appliance consumption, depends upon determination of the accuracy of the hours of usage.

The engineering approach is based on the number, X_{ij}, wattages, W_{ij}, and hours of usage, h_{ij} of the appliance j reported by the sample household $_i$ during the survey. It follows that

$$E_{ij} = X_{ij} \times W_{ij} \times h_{ij} \tag{6.14}$$

Hence the consumption by all the appliances in the ith household is the total consumption by the ith household.

$$E_i = \Sigma_j \, E_{ij} = \Sigma_j \, X_{ij} \times W_{ij} \times h_{ij} \tag{6.15}$$

Thus, engineering estimates of the household consumption can be obtained from the survey data on the appliance stock, wattages and usage. These estimates were calculated for each of the appliance categories in non-AEH and AEH households (Table 6.4). The survey-based estimates show that lighting (40 W and 60 W bulbs) and fans are the major electricity-consuming activities in non-AEH homes whereas in AEH homes, cooking (stove) and water-heating are the major end-uses.

The household consumption estimates were compared with the actual values of the household consumption by regressing one on the other. The resulting correlation coefficient is 0.6, but the standard error of estimate is 61.4 which is nearly 50 per cent of

Table 6.4: Electricity consumption in Bangalore's households (survey-based estimates).

Appliance category	Wattage (kW)	Average appliances/ household		Usage (hr/month)		Consumption (kWh/month)			
						Per appliance		Per household	
		Non-AEH	AEH	non-AEH	AEH	non-AEH	AEH	non-AEH	AEH
25 B	0.025	0.36	1.03	86	120	2.15	3.00	0.77	3.09
40 B	0.040	1.65	2.78	112	90	4.48	3.60	7.39	10.01
60 B	0.060	1.16	2.41	104	80	6.24	4.80	7.24	11.57
100 B	0.100	0.43	0.68	32	30	3.20	3.00	1.38	2.04
Tube	0.040	0.61	2.10	82	120	3.28	4.80	2.00	10.08
EI	0.750	0.43	0.89	8	15	6.00	11.25	2.58	10.01
CF	0.100	0.52	0.65	120	120	12.00	12.00	6.24	7.80
TF	0.017	0.32	0.32	90	110	1.53	1.87	0.49	0.60
GR	0.75		0.2		12		9.00		1.80
MX	0.450	0.11	0.51	3	15	1.35	6.75	0.15	3.44
MT	0.375		0.13		30		11.25		1.46
RD	0.013	0.63	0.84	120	120	1.56	1.56	0.98	1.31
RF	2.0		0.24				30.00		30.00
ST	0.100		0.75		25		50.00		37.50
TV	2.0	0.36	0.70	80	120	8.00	12.00	2.88	8.40
WH			1.01		30		60.00		60.60
Total								32.03	199.71

coefficient is 0.6, but the standard error of estimate is 61.4 which is nearly 50 per cent of the mean consumption (120 kWh/month). This large value of the standard error indicates that there is a tremendous variation between the two values and, therefore, this approach cannot be used. However, for the estimation of future electricity consumption the engineering approach is not useful, therefore, another approach, which can minimise the error and also be useful for estimating future consumption should be considered.

(iii) Appliance stock approach

Since the stock of electrical appliances in a household plays a major role in determining its electricity consumption, it is important to estimate the end-uses through appliance consumption. Instead of measuring this consumption directly by metering (which is an expensive procedure), an econometric approach is adopted here to estimate the appliance consumption.

Let X_{ij} and W_{ij} be the number and wattages of electric appliances of the *j*th category in the *i*th household, and (X_{ij} * W_{ij}) is the electrical load in watts due to this category of appliances. It follows that $L_i = \Sigma_j (X_{ij} * W_{ij})$ is the total load of all the electrical appliances in the *i*th household. It is reasonable to expect that the electrical energy consumption, E_i (in kWh), of households is correlated with their loads L_i (in kW). When the monthly electricity consumption is regressed on the household load, the result is:

$$E_i = 15.4189 + \underset{(0.5038)}{33.8013}\, L_i \qquad (6.16)$$

$$R^2 = 0.92$$

(The number in parentheses represents the standard error)

The conclusion is that a unit increase in load results in an increase of 34 kWh per month in electricity consumption. In addition, the household load due to its appliance stock explains 92 per cent of the variation in electricity consumption. Hence, the stock of electrical appliances explains the bulk of the variation in the electricity consumption of households. Income and other exogenous variables do not seem to dominate the electricity consumption in the short run. Though the pattern of ownership of appliances dominates the short-run demand, the long-run demand is associated with changes in the appliance stock. Therefore, in the long-run, increases in family income would play an important role in increasing the electricity consumption through the acquisition of new appliances, while in the short run, variations in the stock of electrical appliances explain the variations in electricity consumption.

(iv) Appliance census approach

The determination of the contribution of various categories of appliances to the total electricity consumption by regression analysis can be referred to as the *appliance census approach*. The electricity consumption estimates for each appliance category can be made by regressing the quantity of electricity consumed in a household against the number of electrical appliances of different categories. The form of the equation to estimate the monthly electricity consumption of the *j*th appliance category is

$$E_i = \Sigma_j (X_{ij} \times b_j) \qquad (6.17)$$

where b_j, the consumption per appliance of the *j*th category, can be estimated by regressing the electricity consumption E_i on the number of appliances X_{ij} of the *j*th where b_j, the consumption per appliance of the jth category, can be estimated by regressing the electricity consumption E_i on the number of appliances X_{ij} of the jth category in the ith household. Strictly speaking, the coefficient b_j is not the consumption per appliance but the marginal consumption of the jth appliance category, i.e. the increase in consumption resulting from the addition of one appliance of the jth category. Since one expects a linear relation between the consumption and the number of appliances of a particular category, the marginal and the average consumption of the jth appliance category can be taken to be identical.

This approach is used to determine the electricity consumed through different appliances by regressing the average monthly electricity consumption for a combined sample of non-AEH and AEH households on the number of appliances in different categories.

$$\begin{aligned} E = 11.0 &+ \underset{(1.44)}{2.18}\, 25B + \underset{(0.93)}{2.05}\, 40B + \underset{(1.02)}{1.75}\, 60B + \underset{(1.56)}{1.20}\, 100B + \underset{(1.51)}{7.15}\, B + \underset{(3.52)}{9.17}\, EI \\ &+ \underset{(2.10)}{9.34}\, CF + \underset{(2.91)}{3.70}\, TF + \underset{(6.05)}{17.99}\, GR + \underset{(1.74)}{3.78}\, MX + \underset{(3.89)}{18.46}\, MT + \underset{(0.9)}{1.07}\, RD \\ &+ \underset{(3.84)}{17.31}\, RF + \underset{(5.42)}{63.07}\, ST + \underset{(3.24)}{8.18}\, TV + \underset{(3.45)}{74.26}\, WH \end{aligned} \qquad (6.18)$$

$R^2 = 0.93$, and standard error of estimate = 29.3

(Numbers in parentheses represent standard errors)

As already pointed out, the regression coefficients indicate the *marginal* increase in electricity consumption per unit increase in the number of appliances of the jth category. If the increase is linear, then the regression coefficient is also the *average* electrical energy consumed per month through that category of appliance. As expected, all the coefficients are positive. The high value of R^2 (0.93) indicates a good correlation with 93 per cent of the variation in the electricity consumption being explained by the 16 categories of appliances included in the model. The high value of R^2 (0.93) and the low value of the standard error of estimate (29.3) compared with the mean (120 kWh) indicates that the estimates are reliable.

The per appliance consumption is estimated in equation (6.18) for a combined sample of both AEH and non-AEH households. Since the appliance stock varies widely between these two groups, it is necessary to estimate the appliance consumptions separately for AEH and non-AEH households. In AEH households, the per appliance consumption is highest for water-heaters at 56.3 kWh per month followed by stoves at 49.7 kWh per month. The monthly consumption for a refrigerator was 23 kWh whereas a motor consumes 22.1 kWh. In non-AEH households, the ceiling fan is the highest consuming appliance at 10.7 kWh per month. The consumption through other major appliances includes: table fans (5.26 kWh) and electric irons (3.95 kWh). Lights, ceiling fans and television sets are the highest electricity-consuming appliances in non-AEH households (see Table 6.4) whereas in AEH households, water-heaters and electric stoves consume the most electricity (Figure 6.3).

Effect of family size

It would seem reasonable that energy consumption increases as the number of residents in the dwelling increases. This is clearly the case for the consumption of electricity by stoves and water-heaters. The energy consumption of water-heating appliances increases more rapidly than does the consumption of other appliances. This is because hot water is used on an individual basis whereas other appliances are generally used on a joint basis. Consumption through bulbs, grinders, TVs, radios, etc., remains fairly constant as the household size increases. Therefore, the sample is divided into four different family-

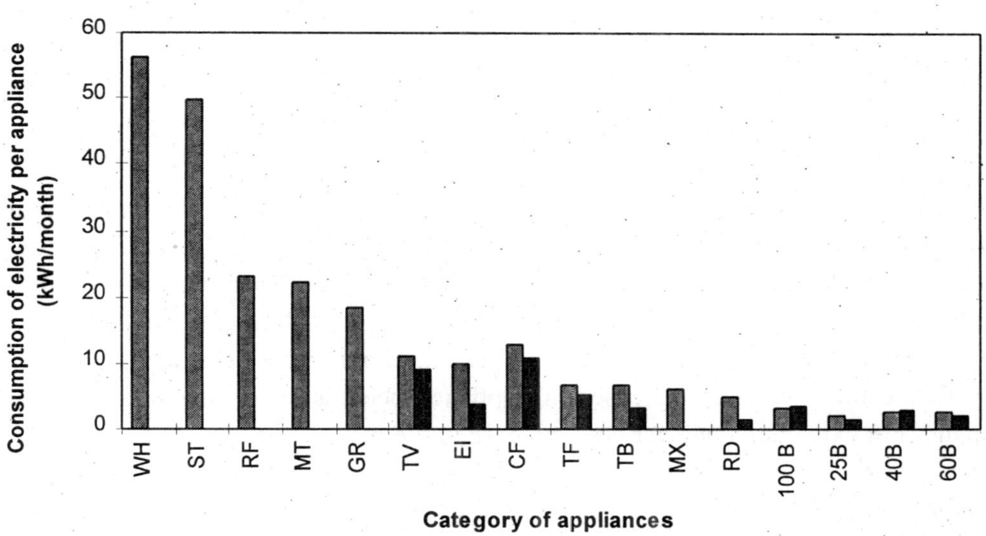

Figure 6.3: Per appliance electricity consumption.

size categories and the consumption through different appliances. The results show that water-heater consumption increased from 45 to 58 kWh/m from category I to IV and the consumption by stoves increased from 41 to 51 kWh/m.

Appliance elasticities of household energy consumption

X_{ij} has been defined as the number of appliances of the jth category in the ith household from which it follows that $X_j = \Sigma_i X_{ij}$ is the number of appliances of the jth category in all the N households. Further, $X_j/N = X_j$ is the average number of appliances of the jth category per household. X_j can also be described as the penetration of the jth category appliances. If the penetration X_j of appliances of the jth category is increased by unity, i.e. one more appliance of this category is added to the appliance stock, then the percentage change in the penetration is $(100 \times 1/X_j)$. In response to this stimulus, the monthly household energy consumption is increased by b_j and the percentage increase in the average monthly household consumption is $(100 \times b_j/E)$ where E is the average monthly electricity consumption of households. The ratio of these two percentage changes is in fact the conventional definition of elasticity. Thus, the appliance elasticity, η, of household electricity consumption is

$$\eta = \frac{\text{Percentagechangeinelectrictyconsumption}}{\text{Percentagechangeinappliancepenetration}}$$
$$= (100 \times b_j)/E]/\ [(100 \times 1)/\ X_j] = (b_j \times X_j)/E \qquad (6.19)$$

For example, in the case of water-heaters in AEH households, $b_j = 56.3$ kWh per household per month, $E = 214.89$ kWh per household per month and the penetration is 1.01. Hence, the water-heater elasticity of electricity consumption is 0.265 (56.3 * 1.01/214.89) which means that a one per cent increase in the penetration of water-heaters results in a 0.265 per cent increase in electricity usage. Table 6.5 gives elasticities of various appliances in both AEH and non-AEH households. These elasticities can be used to estimate the effects of increases in appliance penetration on the future electricity demand. For example, a one per cent increase in the penetration of water-heaters, i.e. an increase in the penetration from 1.01 to 1.02, is likely to lead to an increase in the average household electricity consumption by 0.265 per cent, i.e. 0.00265 * 214.89 = 0.569 kWh per household per month. On this basis, a one per cent increase in the penetration of water-heaters can be estimated to lead to an increase in Bangalore's

Table 6.5: Appliance elasticities.

Appliance category	Appliance-wise consumption (kWh/month)		Appliance penetration		Elasticity	
	AEH	non-AEH	AEH	non-AEH	AEH	non-AEH
25 B	2.14	1.35	1.03	0.36	0.010	0.016
40 B	1.56	2.87	2.78	1.65	0.020	0.160
60 B	1.57	2.17	2.41	1.16	0.018	0.085
100 B	3.17	3.57	0.68	0.43	0.010	0.052
TB	6.63	3.33	2.10	0.61	0.065	0.069
EI	6.90	3.95	0.89	0.43	0.041	0.058
CF	8.76	10.70	0.65	0.52	0.026	0.188
TF	6.80	5.26	0.32	0.32	0.010	
GR	18.50		0.17		0.015	
MX	6.06		0.51		0.014	
MT	22.10		0.13		0.013	
RD	4.96	1.35	0.84	0.11	0.019	0.057
RF	23.00		0.24		0.026	
ST	49.70		0.75		0.173	
TV	11.20	9.03	0.70	0.63	0.036	0.029
WH	56.30		1.01		0.265	

electricity consumption by 1.269 GWh per year. Similar calculations have been carried out to estimate the effects on electricity consumption of increases in the penetration of other appliances (Table 6.6).

It must be stressed here that such projections of electricity demand assume that energy efficiencies will remain constant at the present levels. However, when efficiency improvements are introduced, the appliance elasticities will decrease, thereby the impact of the appliance stock on electricity consumption will get reduced.

End-use Analysis

To estimate the consumption of electricity for different end-uses in the households, the consumption through different appliances should be known. Using the appliance consumption figures, the electrical energy used for different end-uses can be calculated using the following formula.

Table 6.6: Appliance penetration results.

Appliance category	Cons/App (kWh/month)		Elasticity		Increase in consumption (kWh/m)				
					Per household		Bangalore		Total
	non-AEH	AEH	Non-AEH	AEH	non-AEH	AEH	non-AEH	AEH	
WH		56.30		0.265		0.569		105819	105819
ST		49.70		0.173		0.373		69367	69367
TB	3.33	6.63	0.069	0.065	0.023	0.139	6913	25910	32823
EI	3.95	9.90	0.058	0.041	0.019	0.088	5781	16397	22177
TV	9.03	11.20	0.110	0.036	0.037	0.078	11064	14590	25653
CF	10.70	12.76	0.188	0.026	0.063	0.057	18936	10596	29532
RF		23.00		0.026		0.055		10272	10272
40B	2.87	2.56	0.160	0.020	0.054	0.043	16116	8071	24187
RD	1.35	4.96	0.029	0.019	0.010	0.042	2895	7753	10648
60B	2.17	2.57	0.085	0.018	0.029	0.038	8567	7041	15608
GR		18.50		0.015		0.031		5853	5853
MX		6.06		0.014		0.031		5751	5751
MT		22.10		0.013		0.029		5347	5347
25B	1.35	2.14	0.016	0.010	0.006	0.022	1654	4102	5756
TF	5.26	6.80	0.057	0.010	0.019	0.022	5728	4049	9778
100B	2.17	3.17	0.052	0.010	0.017	0.022	5224	4011	9236
Total							82878	304931	387809

$$E_{ik} = \Sigma_i X_{ijk} * b_{ijk} \qquad (6.20)$$

where E_{ik} is the quantity of electricity for the *k*th end-use and X_{ijk} is the number of appliances in the *j*th category for the *m*th end-use in the *i*th household.

Using this equation, the electrical energy consumed for different end-uses in non-AEH and AEH housholds is obtained.

Assuming that the appliance stock in the sample is the same as that of the city, the end-use estimates for residential electric consumption in Bangalore are obtained (Table 6.6). The results show that in the residential sector (i.e. AEH and non-AEH homes), lighting is the major end-use accounting for 23 per cent of the total consumption, followed by water-heating with 22 per cent. Air circulation and cooking

Table 6.7: End-use analysis of electricity in Bangalore's households.

End-use	Appliance category	Consumption per HH		End-use		% of the total	
		non-AEH	AEH	non-AEH	AEH	non-AEH	AEH
Water-heating	WH		56.9		56.9		26.5
Cooking	ST		37.3		37.3		17.3
Lighting	25 B	0.5	2.2	14.1	37.0	49.7	17.2
	40 B	7.1	9.7				
	60 B	2.9	7.5				
	100 B	1.5	3.7				
	Tube	2.0	13.9				
Air-circulation	CF	5.5	23.9	7.1	27.5	25.2	12.8
	TF	1.7	3.7				
Entertainment	TV	3.3	7.8	4.1	12.0	14.5	5.6
	RD	0.9	4.2				
Ironing	EI	1.7	8.8		8.8		4.1
Grinding	GR		3.1		6.2		2.9
	MX		3.1				
Refrigiration	RF		5.5		5.5		2.6
Water-lifting	MT		2.9		2.9		1.3
Others				3.0	20.8	10.6	9.7
Grand total				28.3	214.9	100.0	100.0

share 14 per cent each. The remaining quantity is accounted for by ironing, water-lifting, grinding, etc. (Table 6.7).

Industrial Sector

The industrial sector is the largest end-use consumer of electricity in the total consumption. This sector includes manufacturing, primary metal, chemicals, electronics, food products, etc. The application of electricity in this sector is mainly for process heat, direct heat, electric drives and lighting.

Based on the demand for power, the industrial sector is divided into Low Tension (LT) and High Tension (HT) groups. For LT industries, the connected load is 150 kW per installation, and for HT, the load should be above 150 kW per installation. Even though the number of connections in the industrial sector has increased in the past five years, the electrical energy used per industry decreased substantially. The number of LT industries increased by a factor of 1.5 and the HT industries by 1.06. But the per capita consumption decreased by 3 and 2 per cent in LT and HT industries respectively

(Figure 6.4). This downward trend in the per capita consumption may be due to the power cut imposed by the Electricity Board.

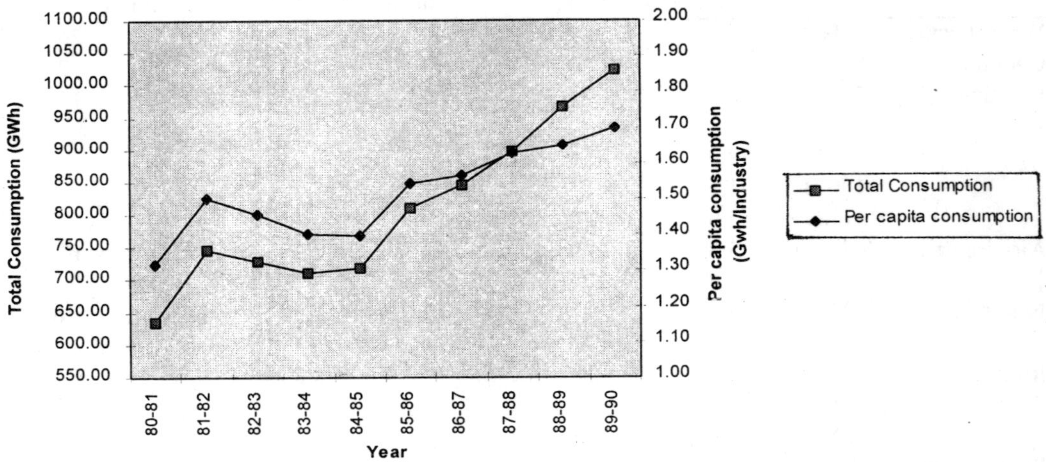

Figure 6.4: Pattern of industrial electricity consumption (1980-90).

The first step for the analysis of electricity utilisation was to identify the structure of consumption so that priorities could be identified. In particular, the choice of conservation options and the flexibility on the consumers' side to shift their load from peak to off-peak periods was to be assessed. Generally, the level of flexibility in the production system would be determined by the dominant mode of production. For example, in a chemical industry producing a given set of products, the scatter of processes in use may vary from fully integrated continuous plants to highly disintegrated plants with a series of batch processes linked with each other through intermediate storage and transfer facilities. All these jobs may not necessarily arrive in a systematic fashion. On the other hand, firms which produce standard products may have a layout to match the product flow through the successive stages of transformation. Their electrical load on the system may follow a more systematic pattern during a day. Thus the extent of diversity of the load pattern within an industrial sub-sector would be determined by the nature and range of products, the dominance of a particular product range within an industry and the method of organisation of production processes.

Electricity is being consumed in the industrial sector for a variety of end-uses. Shaft power and mechanical drive, compressing, pumping and process heating are the main end-uses of electricity with 85 per cent of the total electricity consumption. Of the total electricity consumed (737 GWh) in 1984-85, 26 per cent goes for shaft power, followed by compression with 24 per cent. Pumping and process heating share 19 and 18 per cent respectively. The other major end-uses are blowing (11 per cent) and lighting (4 per cent). (Figure 6.5)

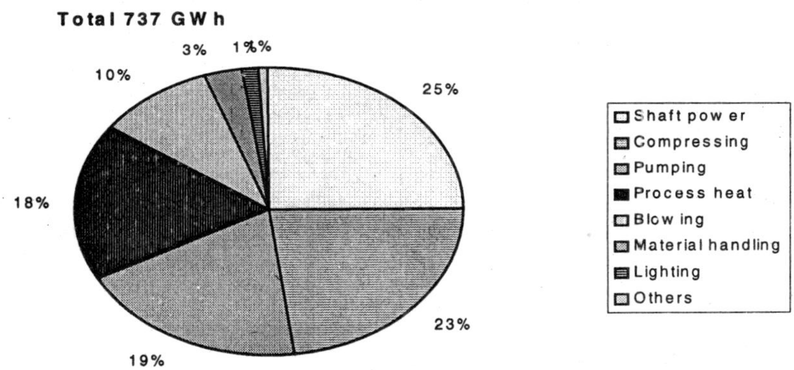

Figure 6.5: End-use analysis of electricity in the industrial sector.

Commercial Sector

The commercial sector consists of a diverse group of consumers ranging from offices, shops, lodgings and restaurants to railway stations, hospitals and movie theaters. Lighting, cooking, grinding and air-cooling are the main uses of electricity in this sector. The number of consumers doubled between 1980 and 1990 and the consumption increased from 42.5 to 65.0 GWh. Thus, the per-connection consumption decreased from 803 to 594 kWh from 1980 to 1990. This decrease may be due to the substitution of energy carriers and the imposition of power cuts. (Figure 6.6)

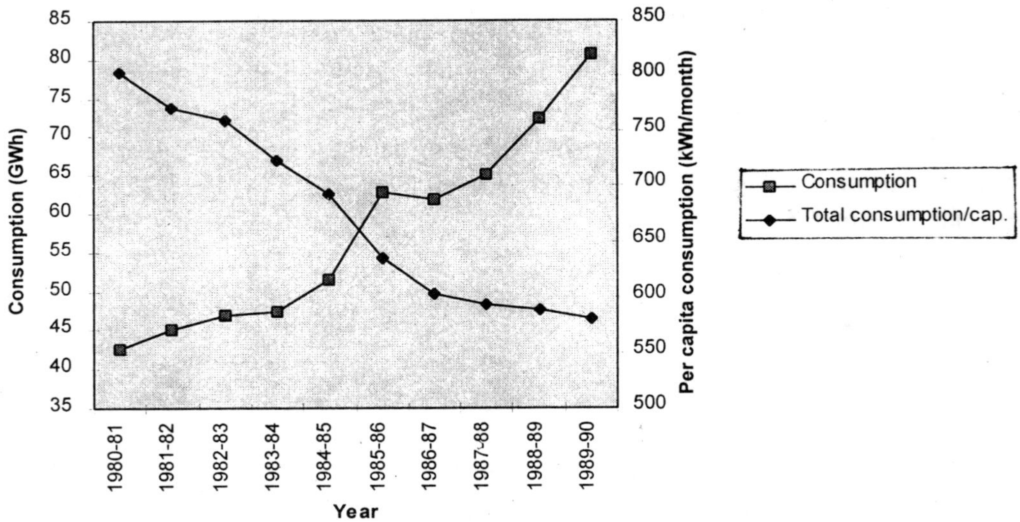

Figure 6.6: Pattern of electricity consumption in the commercial sector.

In the commercial sector, lighting, with 75 per cent, is the major end-use followed by air-circulation with 18 per cent. Cooking, water-heating, etc, account for the remaining quantity. (A. K. N. Reddy, *et al.*, 1988)

End-use Analysis of Electricity in Bangalore

Electricity is used for many applications as has already been mentioned. Lighting, with 16.5 per cent of the total consumption, is the main end-use of electricity followed by pumping with 14.4 per cent and shaft power with 12.8 per cent. Compressing, process heat and water-heating with around 10 per cent each and cooking, air-circulation, grinding and blowing with 5 per cent each are the remaining end-uses. (Table 6.8)

Discussion: Electricity Conservation and Demand Side Management

Demand side management involves efficiency improvements, formulation of pricing policies, good housekeeping practices, load management strategies, etc. In recent years, it has emerged as one of the major concepts which is technically feasible and economically promising. A significant reduction in energy bills of the consumers can

Table 6.8: End-use analysis of electricity in Bangalore (GWh) (1989-90).

End-use	Residential	Industrial	Commercial	Street lighting	Total	% of total
Lighting	133.32	10.00	39.00	25	209.47	15.64
Shaft power		183.30			183.30	13.69
Compressing		169.60			169.60	12.67
Pumping		140.30			140.30	10.48
Process heat		131.00			131.00	9.78
Water-heating	126.96		1.00		127.96	9.56
Air-circulation	87.12		9.50		96.62	7.22
Cooking	83.16		1.00		84.16	6.28
Blowing		74.70			74.70	5.58
Entertainment	41.52				41.52	3.10
Ironing	25.80				25.80	1.93
Material handling		22.00			22.00	1.64
Grinding	13.92				13.92	1.04
Refrigeration	12.36				12.36	0.92
Water-lifting	6.36				6.36	0.47
Others	55.44				55.44	4.14
Total	530.52	730.90	50.50	24.68	1339.07	100.00

occur, if these measures are adopted. The payback period for many of the conservation programmes is less than two years (Reddy *et al.*, 1991; and Parikh *et al.*, 1994).

Despite these potentially significant benefits to the economy and their proven success in several countries, conservation programmes are yet to be undertaken on a significant scale in India. There are several reasons for this. Firstly, a proper understanding of energy conservation and its importance are still unclear to most of the utilities as well as to the consumers. Also, the main concern of the utilities is revenue generation and, in some cases, there are doubts that introduction of conservation measures may affect the revenues adversely. Secondly, there is a misunderstanding about energy conservation as such. Conservation, as understood by many, does not mean the usage of less energy at the cost of output. With the increase of device efficiency one can get the same or more energy services with less energy input and, therefore, without any loss of the output. And finally, the major impediment is the lack of clear policies on energy pricing.

There are many proven technologies that can be considered for energy conservation (Parikh *et al.*, 1994). These measures are referred to as Demand Side

Management Options (DSMO) which emphasise the amount of energy saved or the capacity the option reduces. If the technology is cost-effective, then it can be considered for DSMO. This cost-effectiveness can be defined as the benefit (cost saved due to avoiding new capacity by implementing DSMO) to cost (cost of DSM alternative) ratio.

The various DSMOs that have a significant conservation potential are: (i) energy efficient motors, (ii) variable speed drives, (iii) replacement of incandescent bulbs with compact fluoroscent lamps, (iv) replacement of magnetic ballasts with electronic ballasts, and (v) replacing electric water-heaters with solar water-heaters. Thus, once such DSMOs have been characterised in terms of their cost, savings, market penetration, etc., the next step is to evaluate their cost-effectiveness against supply-side resources which they are intended to replace. Here, the costs and benefits of DSMO are compared to the supply-side alternatives, i.e. marginal costs of power generation.

Motor systems

Electric motors are used for (i) the movement of material: pumping, blowing, and solids conveying, (ii) mechanical processes: pumping and fan, and (iii) cooling.

In the industrial sector, motor-driven end-uses account for nearly 70 per cent of the total electricity consumption, thereby gaining considerable attention as a potential load modification potential. For end-use application, characterised by long and constant load conditions, a high-efficiency motor can be used instead of a standard motor to improve efficiency and reduce electricity consumption. Generally, motors in India are always oversized and, therefore, operate at part load. For optimum utilisation, the important changes to be carried out in the motor are: (i) power factor correction, (ii) phase-balancing, and (iii) rewinding without burning the stator core. Improved design and materials result in improved motors (efficiency is 3-5 per cent higher) known as energy efficient motors (EEM's). In our analysis, efficiencies of existing standard motors of reputed manufacturers like Kirloskar, Siemens, etc., are considered. The actual efficiencies (of the standard motors) are likely to be lower as many industries use motors supplied by small scale units.

Fluids driven by pumps, fans and blowers are usually regulated by throttling. This is an inefficient method of control and results in significant energy losses. A more efficient control method is to control the speed of the motor using a variable speed drive (VSD). Variable speed drives permit continuous regulation of motor speeds which can lead to substantial electricity savings, particularly, when there is partial load operation during a large proportion of the duty cycle. Variable speed drives have significant scope in the paper, chemical, fertiliser, pharmaceutical and cement industries. These VSDs are widely regarded as a promising technology with significant electricity savings potential. In addition, smooth motor startups and shutdowns, prolong motor life and process control, and because of their small size, have tremendous retrofit potential. All these measures may reduce the electricity consumption by 25 - 30 per cent (Parikh *et al*, 1994).

Lighting Systems

On the basis of total energy consumption, lighting is a relatively minor end-use accounting for only 4 per cent of the total demand. However, since lighting is one of the few end-uses for which no practical non-electric alternative exists, it is important to study the conservation aspects. The options that are considered here are: (i) replacement of incandescent bulbs by compact fluorescent lamps, and (ii) replacement of magnetic ballast with electronic ballast.

Incandescent bulbs are commonly used in domestic and commercial sectors. Though incandescent bulbs account for only about 4 per cent of the total consumption, it may be worthwhile to replace them by compact fluorescent lamps leading to energy savings of the extent of 75 per cent. New compact fluorescent lamps last about eight to ten times longer than incandescent bulbs and are four times as efficacious. Compact fluorescent bulbs are now available in the country. An incandescent lamp with a rating of 60 W has an output of 700 lumens, a life of 1,000 hours and is priced at Rs.9.60 (1992-93 values). A 11 W compact fluorescent lamp has an output of 900 lumens, a life of 8,000 hours and is priced at Rs.200. The auxiliary consumption of the ballast is 3 W. The cost of the ballast is Rs. 150/- and the life of the ballast is 10 years.

Fluorescent tubes are widely used in residential, commercial and industrial establishments. Each tubelight fixture has a ballast which provides a high voltage to initiate the discharge and then limit the current. A conventional magnetic ballast for a 40W rating fixture consumes about 12W of power. Instead of a magnetic ballast, it is possible to opt for an electronic ballast which draws only 1-3W.

In the residential sector, water-heating is the largest single end-use that utilises nearly 24 per cent of the total consumption. As the appliance elasticities indicate, a one per cent increase in water-heaters in Bangalore can increase the electricity consumption by 3,527 kWh per day. This large quantity of precious energy can be saved by substituting solar water-heaters for electrical water-heaters. A flat bed solar panel of area of 2 m^2 is expected to provide approximately 2.5 ltr per minute of hot water at 60^0 C. (Anon, 1990). Since the temperature required to heat water is around 45^0 C, which can easily be met by solar energy, installation of solar water heaters is an important consideration.

The technical characteristics and costs of various DSM options are obtained from manufacturers based on which the energy savings and the peak demand reduction were quantified (Table 6.9).

In the analysis, a peak coincidence factor (0.7 to 0.9) was considered. The percentage of unwilling customers (who are not interested in investing in DSMOs) varies from 10 to 30. The DSM options considered here have been evaluated at the tariff structure of Rs 2.0/kWh for energy charge and Rs 100/kW for demand charge (1994-95 tariff) with an escalation rate of 10 per cent per year. It is assumed that the electricity consumption will grow at the rate of 7 per cent in the first year and reduce gradually to 5 per cent at the end of the 15 year period.

Energy and Demand Savings

Once a DSM option has been characterised in terms of its savings, market penetration, etc., the next step is to evaluate its effectiveness through demand reduction and energy savings.

The energy savings and peak demand reduction varied from one DSM Option to another. The total energy savings per annum will be 1,148 GWh and the peak demand

Table 6.9: Details regarding various DSM options.

DSM option	Cost	Energy/ demand savings (% higher)	Life (years)	Payback period (years)		Rate of return (%)
				Simple payback	Discounted payback	
Energy efficient motors	30% higher than standard one	10	15	2.5	2.9	64.2
Variable speed drives	Rs.9,500 to 11,000	300	15	2.7	3.0	44.5
Compact fluoroscent lamps	35 times higher		8 times higher	0.5	0.7	200.0
Electronic ballasts	100% higher		2			
Solar water-heaters	5 times higher	4 times	10	3.2	3.2	41.1

reduction will be 258 MW. The contribution from different programmes indicates that an effficient motor system is the most promising option contributing 436 GWh of the total savings. This is followed by CF with 258 GWh (23.8 per cent). The energy savings from electronic ballasts will be 183 GWh (19.6 per cent). Substituting SWH for electrical water-heating saves 160 GWh. (Table 6.10)

There will be substantial increments in electricity consumption if the DSM programmes are not implemented. By 2000 A.D., the electricity consumption will increase significantly. If the DSM programmes are implemented, there will be a substantial reduction in the peak demand and electricity consumption. This is depicted in Figure 6.7.

Table 6.10: Peak demand and energy savings through efficiency improvements and carrier substitution.

End-use	Sector	Savings	
		GWh/year	MW
CF	Non-AEH	145.6	
	AEH	110.9	
	LT	3.1	
	HT	8.3	
	Commercial	14.2	
	Total	267.9	82.9
Ballasts	Non-AEH	56.1	
	AEH	41.7	
	LT	19.4	
	HT	48.0	
	Commercial	17.9	
	Total	183.1	41.3
Motors	LT	4.4	
	HT	432.2	
	Total	436.6	98.3
SWH	AEH	160.0	36.1
	Grand total	1147.6	258.5

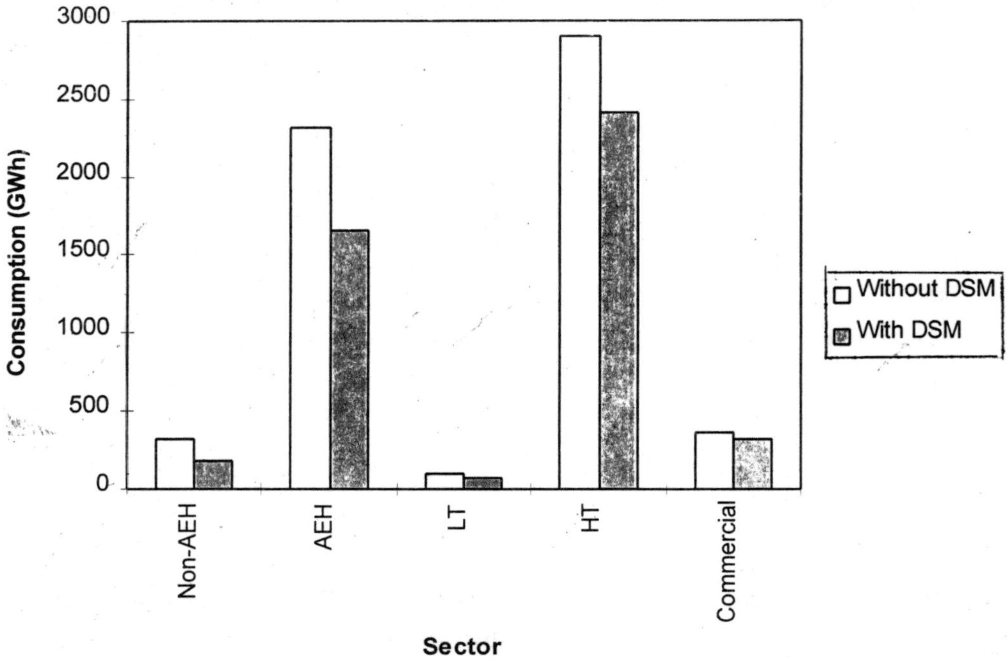

Figure 6.7: Effect of DSM on electricity demand by 2000 AD.

Impact on environment

The implementation of DSM programmes helps to reduce pollution over the baseline. The burning of coal in power plants pollutes the environment through emissions of carbon dioxide, sulphur dioxide, nitric oxides and flyash. The calorific value of the Indian coal is also decreasing rapidly, thereby increasing its consumption per unit of generation. Therefore, the energy savings from DSM programmes will reduce the emissions significantly. The quantitative assessment of various emissions associated with power generation in thermal plants has been carried out using the norms from Chattopadhay and Parikh (1993). Table 6.11 provides details regarding the emission reduction. As can be seen from the table, the quantity of emission reduction in CO_2 will be 1.1 million tonnes by the year 2020 A.D. The SO_X and NO_X reductions will be 3,000 and 5,000 tonnes respectively by 2020 A.D. The quantity of flyash that can be reduced will be 4,500 tonnes.

The above results clearly demonstrate that investments in DSMO can avoid or minimise the necessity to construct new power plants (supply side options). This can be done by integrating the DSMOs with the supply-side process. By doing so, the investment plans, already drafted by the KEB to meet the anticipated load, can be altered. The cost of DSMOs is significantly lower than the capital investment in power generation required by the utility, which ranges between Rs.30,000 and 40,000/kW. This suggests that through investments in these DSMOs one can reduce the peak demand and electricity consumption. This can be done by integrating the DSMOs with the supply-side resources. By doing this, the investment plans, already drafted by KEB to meet the anticipated load, can be altered. For preparing a least-cost plan for the KEB, the DSMOs can also be considered as supply resources and can be treated on a par with other resources. In order to make this DSM plan a least-cost one, supply and DSM inputs must be compared on an equivalent basis (EPRI, 1987). For this, some adjustments should be made in the final calculations. Generally, supply-side resources are evaluated at the generation level. In the case of DSMOs, they are installed on the customer side of the meter. Therefore, DSM options not only reduce the need for generation but also the need for necessary transmission and distribution (T & D) facilities. Therefore, the transmission and distribution loss factor (0.2) can be credited to the demand reduction, energy savings and fuel and capacity savings.

The results of this study show that DSMOs are considerably cheaper than building new power plants. Then, why is it necessary to have incentives and subsidies, if the measures are so cheap? The reason is that customers do not compare their costs with those of building new power plants but with other investment opportunities which may

Table 6.11: Emission reductions through the implementation of DSM options.

DSM option	Energy savings (GWh)	Emission reduction (000 tonnes)			
		CO_2	SO_x	NO_x	Flyash
CF	268	281	0.77	1.30	1.15
Electronic ballast	183	192	0.53	0.89	0.79
Motors	437	458	1.26	2.11	1.88
SWH	160	168	0.46	0.77	0.69
Total	1048	1100	3.01	5.07	4.50

give a higher rate of returns. Their borrowing rates can range from 18 to 25 per cent whereas utilities (or social discount rates) pay lower rates. Customers will not purchase more efficient equipment, if costly, and more power plants will be required to add more industrial customers to the utility. Electricity rates for customers are already high. Often they cross-subsidise for agricultural consumption. Hence the need for DSM programmes with cost-sharing. If, due to DSM, more customers are added with the same power capacity, then when the costs go up, the rise is shared by a larger number of customers. Hence, either the government or the electricity boards, or both, should share a part of the costs of these energy-efficient appliances and help the consumers to substitute old, inefficient appliances with new and efficient ones. This cost funded by the utility or manufacturers or some other agency can be recovered from the consumers in instalments along with their monthly electricity bills. The consumer will recover his investment through savings in electricity consumption in a reasonable period (less than two years), leading to an overall improvement in the availability of power.

Use of electricity pricing as a signal to consumers

The present electricity tariff is much lower than the incremental cost of new capacity, i.e. long-run marginal cost. Therefore, appropriate electricity tariff for AEH/industrial users would be a powerful incentive for conservation. It should reflect all external costs associated with energy generation, transmission and distribution including environmental degradation. Such a price structure would reward efficient use patterns and discourage inefficient use patterns. The fluctuations in the daily load curve of the Karnataka State Electricity System emphasise the importance of load flattening. This can be achieved by imposing an appropriate tariff structure such as time-of-use tariffs. Relating charges for electricity to the time of day when it is used, particularly vis-a-vis peak and off-peak system demand periods, is a potent DSM measure, extensively used overseas. Considerably higher demand and energy charges during the peak periods discourage it and promote transfer of some of the load to off-peak periods. Along with the reduced energy charges for off-peak periods, increased rates for peak periods are necessary to create sufficient margins for load shifting. The tariffs need to deal with at least eight-hour off-peak duration and not more than three to four hours of the peak period (Parikh *et al.*, 1995). Only a clearly indicated long term pricing policy will bring about the desired changes.

Importance of new industries/all-electric homes

While most DSM programmes can and do benefit the immediate and short-term power scenarios, the significant growth rate of demand and consumption magnifies long-term benefits in absolute terms. This accelerates the necessity of adoption of such programmes.

In the total electricity consumption of Bangalore, the residential sector consumes about 45 per cent. With the introduction of AEH connections, its consumption has increased tremendously in recent years. For instance, in 1990-91, Bangalore consumed 26.8 per cent of Karnataka's electricity and AEH consumers accounted for 46 per cent of Bangalore's consumption (Anon, 1991). Thus, Bangalore's AEH consumers used 11.1 per cent of the State's electricity consumption. Within these homes, water-heating and cooking are the main end-uses. Since AEH consumption permits the use of appliances that draw as much as 15 amperes, it automatically enables the household to reach a higher consumption level. In addition, since the marginal price of electricity to an AEH consumer does not increase with consumption, the consumer is encouraged to use appliances such as refrigerators and grinders. Thus, the AEH can be considered to be an important factor in the increase of consumption of electricity in the residential sector. In the total electricity consumption the industrial sector consumes about 35 per cent (1994-95 figures) and the revenue from this sector is more than 50 per cent. Over a 25-year period, with an assumed growth rate of 7 per cent, the major contributor to energy consumption will be the new industries. Thus, industrial consumption can also be considered to be an important factor in the increase of consumption of electricity. It will be worthwhile, therefore, to fix norms for new industries. Before providing a licence or connection to a new industry the energy plans of the new units should be examined and approved. Databases and manuals for efficient technologies need to be prepared and made available, so that industries make the right choice.

Testing facilities and standards

It is essential to incorporate data on the maximum energy consumption for equipment like motors, fans, blowers and bulbs, in the specifications. Norms for different products should be drawn up and testing of energy efficiency of electrical equipment made mandatory before granting the BIS (Bureau of Indian Standards) certificate. Data on the efficiency of the equipment should be included in its nameplate.

There are some who feel that a market-based approach alone does not deliver the goods and regulations are needed. This could be the case especially for standards of end-use equipment. Periodic audits of energy-intensive industries need to be mandatory. Reporting energy use in the annual reports is already mandatory but the format has not been given an adequate thought. It is difficult to judge whether the energy efficiency of the enterprise has gone up or not from the reported data. Giving consumption norms is another possible guideline.

Formation of a nodal DSM agency

A consortium consisting of representatives from utilities, financial institutions, equipment manufacturers and industries could implement DSMOs. It is rather important that this nodal agency should be formed to function on a self-sufficient or even a commercial basis, deriving its financial sustenance from savings through DSM options. In the USA, this type of agency bids commercially to deliver specified amounts of demand and energy savings to utilities. In India, this agency can have a wider role in response to the above needs.

7

THE FLOW OF ENERGY IN AN URBAN SOCIETY

Nature has bestowed upon us an inquisitive disposition....
Our vision opens up a path for its investigation and lays the foundation of truth so that our research may pass from revealed to hidden things.

– Anon

The growth of an urban society depends, to a great extent, on the quantity of energy it consumes. The bigger the city, the greater its dependence on energy particularly on high-grade carriers like electricity and petroleum products. From collecting biomass pickings to owning piped-gas homes, man has come a long way. His efforts in evolving an energy cycle, converting energy from less-desired forms to more convenient and efficient ones, are remarkable.

In the early stages, man's needs were limited to the use of firewood for heating. As times changed, his needs increased thereby increasing his energy requirements manifold. The present-day man consumes more than 100 times the energy consumed by the man of agricultural civilisation. The explosive growth of urban areas has led to a sharp increase in the consumption of petroleum products and electricity. The rising urban demand for energy has to be addressed as a part of national development plans but this cannot be done without an understanding of the flow (from supply to end-use service) of various fuels and energy carriers and their utilisation in urban conglomerations. The study of the coexistence and utilisation of several energy carriers in a stratified city is also an interesting one. All the energy that is being supplied to the city is not converted into useful energy because of the low efficiency of utilisation. From resource to end-use, the energy utilisation is only about 50 per cent! The remaining goes waste.

This chapter is aimed at providing an insight into the energy sector of Bangalore. It focusses attention on the *fuel cycle* of supply, transportation/transmission and distribution of energy carriers viz., firewood, charcoal, coal, electricity, LPG and petroleum products like kerosene, petrol, diesel and fuel oil. The consumption pattern of these energy carriers for various end-uses such as cooking, water-heating, lighting, process heating, etc., in the residential, industrial, commercial and transport sectors is discussed. Finally, the implications of energy consumption on transportation, forests, environment and the economy as a whole are discussed and policy measures suggested. Figure 7.1 shows the flow of energy in Bangalore from supply to end-use.

Energy Supply

If the supply aspect of the energy is considered, out of the 45,240 TJ (tera joules or 10^{12} Joules) of energy consumed in 1989-90 in Bangalore, 2,956 TJ (6.5 per cent) came from within a distance of 100 km (in the form of firewood) and 3,208 TJ (7.1 per cent) from a distance greater than 1,000 km (in the form of coal and LPG). The balance came from between 100 and 1,000 km (petroleum products, firewood, charcoal and electricity). (Figure 7.2)

Suppliers

In 1989-90, Bangalore's energy was supplied by the Central Government to the extent of 27,345 TJ (60.4 per cent of the total) in the form of coal, LPG and petroleum products. The Government of Karnataka supplied 8,324 TJ of energy (19.2 per cent) in the form of firewood and electricity, and private contractors supplied 8,545 TJ of energy (18.9 per cent) in the form of firewood and charcoal. The remaining quantity was supplied by the Government of Andhra Pradesh in the form of coal (Figure 7.3).

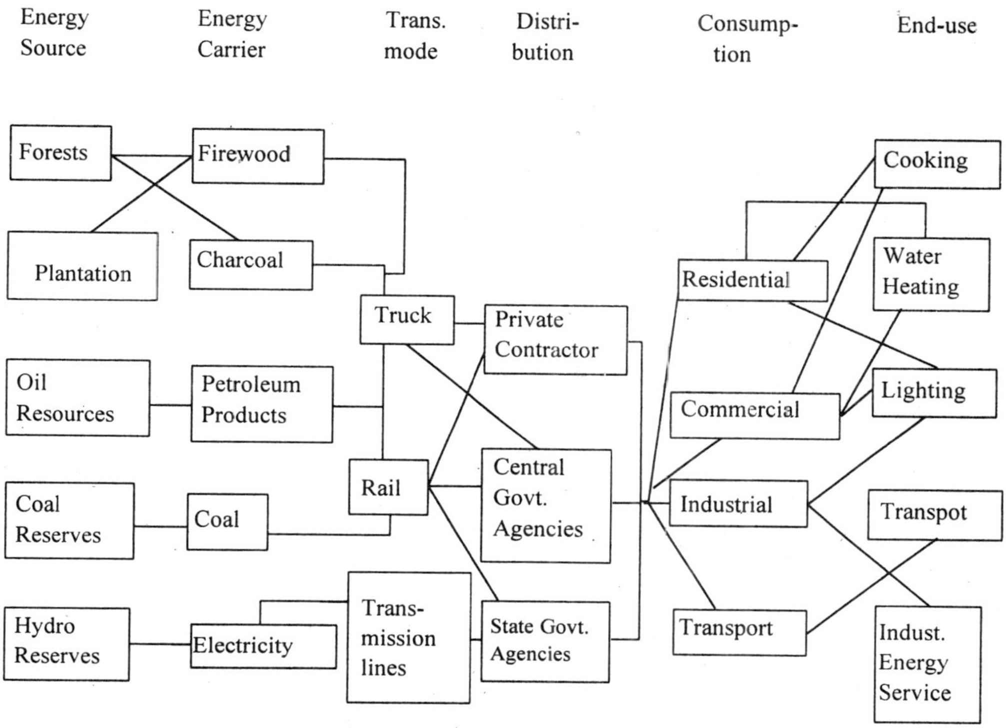

Figure 7.1: Flow of energy in Bangalore from supply to end-use.

Transportation

Railway wagons and trucks were the main modes for the transportation of energy carriers to Bangalore. Of the 45,240 TJ of energy transported to Bangalore, during 1989-90, 24,572 TJ (54.3 per cent) was transported through railways (in the form of petroleum products, firewood, and coal), 12,112 TJ (26.8 per cent) by trucks (in the form of LPG, firewood and charcoal). 8,324 TJ (18.5 per cent) of electrical energy was transmitted through the grid (Figure 7.4).

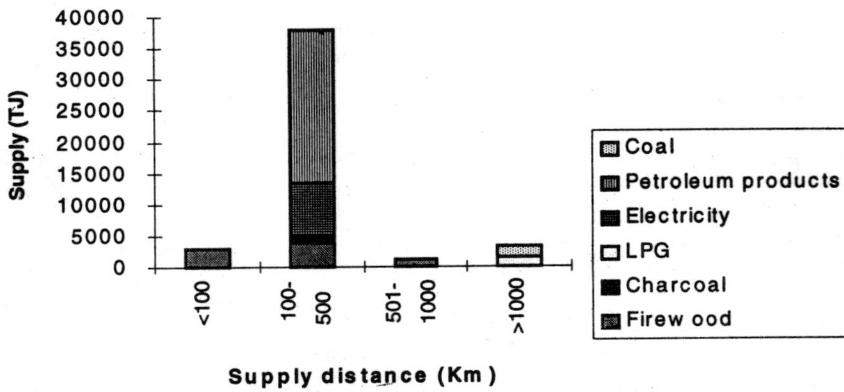

Figure 7.2: Supply of various energy carriers.

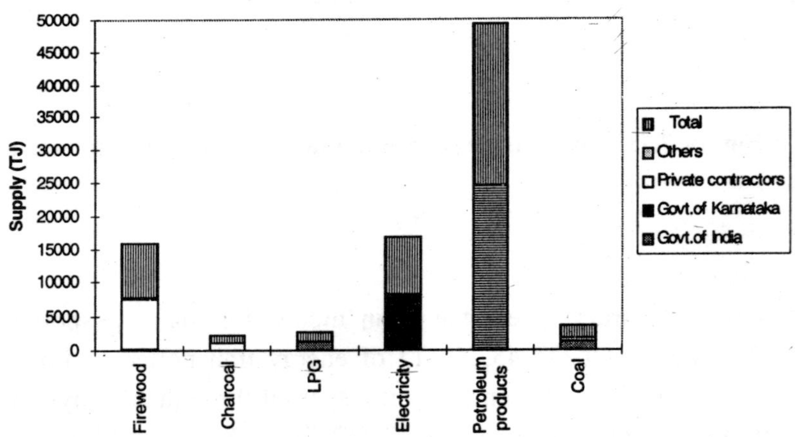

Figure 7.3: Suppliers of various energy carriers.

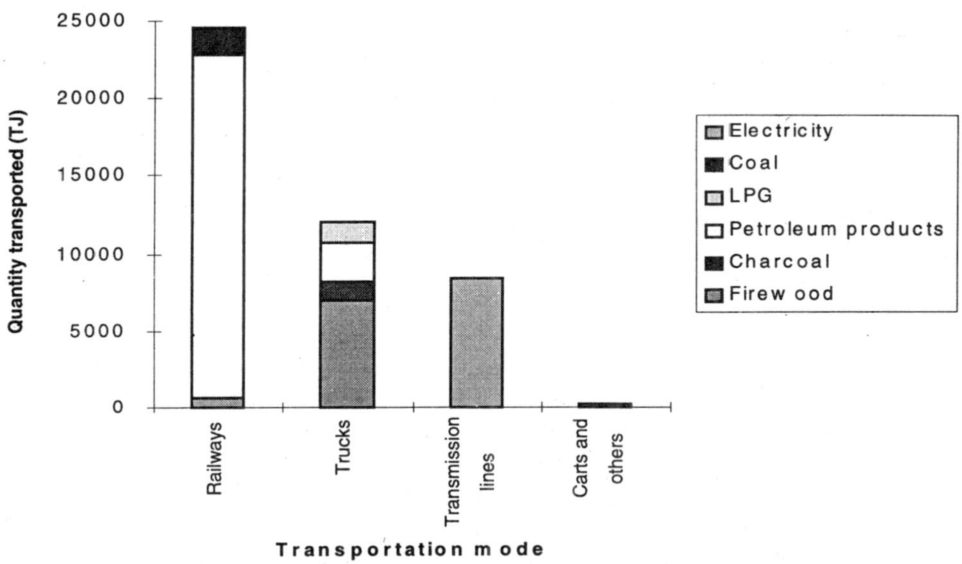

Figure 7.4: Transportation modes of various energy carriers.

Distribution

The Oil Co-ordination Committee distributed 24,603 TJ (57.5 per cent) of energy in the form of petroleum products and LPG, the Karnataka Electricity Board, 8,364 TJ (18.5 per cent) as electricity, Coal India Ltd., 1,788 TJ (4.0 per cent) in the form of coal, and private contractors, 9,065 TJ (20.0 per cent) in the form of firewood and charcoal (Figure 7.5).

Trends in Energy Consumption

Between 1980 and 1990, the total energy consumption increased by a factor of 1.3 with an annual growth rate of 4.27 per cent per annum. The shares of various energy carriers in the energy mix varied considerably during this period. Although, the end-use consumption of firewood and charcoal increased marginally from 8,207 to 9,065 TJ, their share in the total energy consumption decreased considerably, i.e. from 27.6 to 20.1 per cent. In the case of kerosene, both the quantity and the share increased from 2,966 to

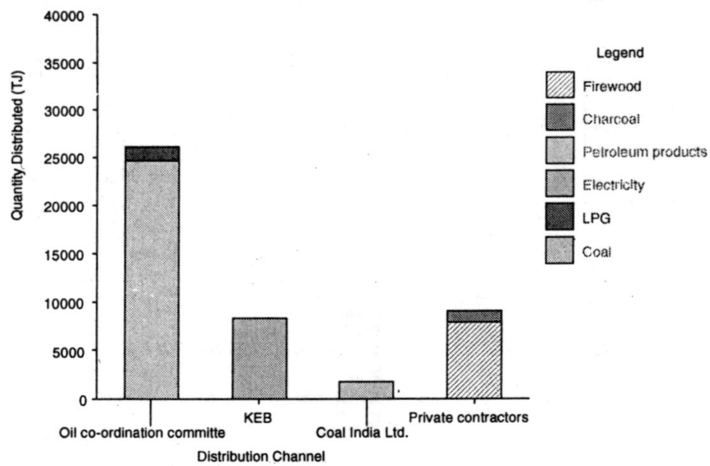

Figure 7.5: Distribution channels of energy carriers.

6,452 TJ and 10 to 14 per cent respectively. The consumption and share of LPG increased significantly from 422 to 1,420 TJ (1.4 to 3.1 per cent). The consumption of electricity has increased from 3,697 to 8,364 TJ during 1980 to 1990 and its share in the total consumption increased from 13 to 18.5 per cent. The consumption of all the petroleum products (petrol, diesel, HSD and LDO) continued to increase despite a substantial increase in their prices. Their consumption rose from 9,029 to 28,151 TJ and their share in the total consumption increased from 30.4 to 40.2 per cent between 1980 and 1990. The use of coal declined sharply from 5,398 to 1,790 TJ and its share fell from 18.2 to 4 per cent (Table 7.1).

In the total energy consumption, the bulk of the energy is used by the residential sector (Figure 7.6). During 1989-90, for instance, Bangalore consumed 45,240 TJ of energy. This energy was utilised by the various sectors in the following way: residential 17,248 TJ (38.1 per cent), industrial 13,240 TJ (29 per cent), transport 12,242 TJ (27.1 per cent) and commercial 1,578 TJ (3.5 per cent).

Table 7.1: Trends in energy carrier consumption (1980-90).

Energy carrier	1980-81	1981-82	1982-83	1983-84	1984-85	1985-86	1986-87	1987-88	1988-89	1989-90	Growth rate (per cent)
Firewood	6604(22.2)	6432(20.9)	6716(21.4)	6891(21.5)	7300(21.8)	7310(20.6)	7450(19.8)	7640(19.2)	7789(18.5)	7861(17.4)	2.10
Charcoal	1603(5.4)	1564(5.1)	1510 (4.8)	1490 (4.7)	461(4.4)	1425(4.0)	1390(3.7)	1360(3.4)	1285(3.1)	1204(2.7)	-2.32
Kerosene	2966(10.0)	3405(11.1)	3771(12.0)	3935(12.2)	4166(12.4)	4515(12.7)	4928(13.1)	5558(14.0)	5948(14.1)	6452(14.3)	9.39
LPG	422(1.4)	475(1.5)	574(1.8)	730 (2.3)	869(2.6)	946(2.7)	1031(2.7)	1147(2.9)	1279(3.0)	1420(3.1)	15.35
Electricity	3697(12.4)	4316(14.0)	4525(14.4)	4852(15.2)	5126(15.3)	6173(17.4)	6696(17.8)	7032(17.7)	7586(18.0)	8364(18.5)	9.62
MS	1893(6.4)	1989(6.5)	2060(6.6)	2159(6.7)	2348(7.0)	2484(7.0)	2622(7.0)	2827(7.1)	3234(7.7)	3610(8.0)	5.90
HSD	4635(15.6)	5066(16.5)	5212(16.60)	5904(18.4)	6204(18.5)	6896(19.4)	7759(20.6)	8411(21.2)	9465(22.4)	10512(23.2)	8.89
LDO	413(1.4)	409(1.3)	401(1.3)	414(1.3)	428(1.3)	501(1.4)	546(1.5)	593(1.5)	645(1.5)	702(1.6)	5.30
FO	2088(7.0)	2230(7.3)	2133(6.8)	2294(7.2)	2374(7.1)	2518(7.1)	2660(7.1)	2811(7.1)	3019(7.2)	3327(7.4)	4.34
Coal	5398(18.2)	4879(15.9)	4487(14.3)	3359(10.5)	3224(9.6)	2777(7.8)	2549(6.8)	2364(6.0)	1942(4.6)	1788(4.0)	-9.74
Total	29719(100.0)	30765(100.0)	31389(100.0)	32028(100.0)	33500(100.0)	35545(100.0)	37631(100.0)	39743(100.0)	42192(100.0)	45240(100.0)	4.27

Note: Numbers in parentheses represent percentages.

Residential Energy Consumption

Trends in consumption

From 1980 to 1990, the total energy consumption of the residential sector increased by a factor of 1.3 while the increase in the number of households was 1.5 (Anon, 1985) which indicates that the per household consumption decreased. This decrease may be due to the changing mix of energy carriers. Many households that used to depend on wood-based fuels have shifted to modern energy carriers like LPG and electricity (Figure 7.7).

Thus, wood-based fuels (firewood and charcoal), which constituted 60 per cent of the total residential energy consumption in 1980, accounted for only 40 per cent in 1990. The share of kerosene in the total energy consumption stood around 30 per cent. LPG had increased its share from 3 to 6 per cent and electricity had almost doubled its share from 11 to 22.6 per cent (Table 7.2).

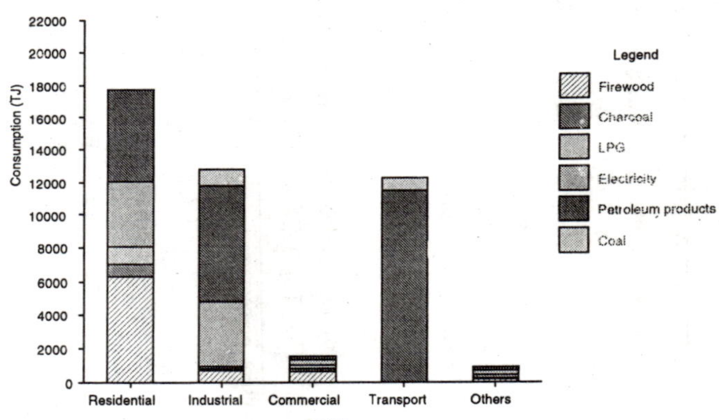

Figure 7.6: Sectoral energy consumption.

Table 7.2: Residential energy consumption (1980-90).

Energy carrier	1980-81	1981-82	1982-83	1983-84	1984-85	1985-86	1986-87	1987-88	1988-89	1989-90	Growth rate (per cent)
Firewood	5151(48.8)	5146(47.9)	5267(45.9)	5320(43.5)	5723(43.7)	5966(41.9)	6015(39.9)	6112(38.5)	6231(37.4)	6289(35.7)	2.91
Charcoal	1083(10.3)	1056(9.8)	1015(8.8)	1063(8.7)	988(7.5)	912(6.4)	870(5.8)	802(5.1)	758(4.5)	710(4.1)	-4.12
Kerosene	2875(27.2)	2821(26.3)	3176(27.7)	3338(27.3)	3589(27.4)	3890(27.3)	4249(28.2)	4790(30.1)	5126(30.8)	5560(31.6)	7.50
LPG	309(2.9)	339(3.2)	420(3.7)	532(4.4)	648(4.9)	699(4.9)	763(5.1)	851(5.4)	949(5.7)	1054(6.0)	15.32
Electricity	1148(10.9)	1382(12.9)	1609(14.0)	1980(16.2)	2159(16.5)	2790(19.6)	3190(21.1)	3337(21.0)	3600(21.6)	3970(22.6)	15.94
Total	10566(100)	10744(100)	11487(100)	12233(100)	13107(100)	14257(100)	15087(100)	15892(100)	16664(100)	17583(100)	6.28

Note: Numbers in parentheses represent percentages.

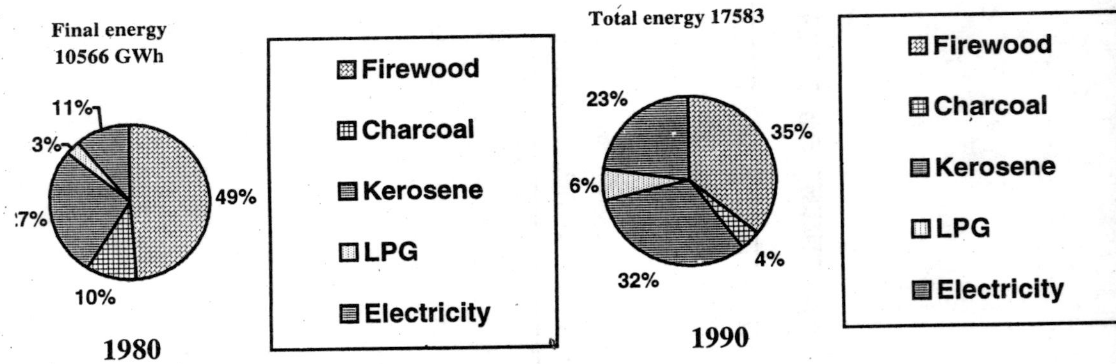

Energy consumption (1980) **Energy consumption (1990)**

Figure 7.7: Change in household energy consumption (1980 to 1990).

The change in consumption of energy carriers was supported by the household survey data of Bangalore. The data indicate that there is a variation in the contribution of different energy carriers to the energy mix of different income groups for cooking and water-heating (Figure 7.8).

Carrier Utilisation index

The household data was used to calculate the utilisation indices of various energy carriers. The utilisation index defines the fraction of households in a particular income group which use a particular energy carrier for a particular end-use. Values of the utilisation index show that the low-income groups (I and II) depend mainly on firewood and agricultural wastes. The middle income groups (III to V) depend on kerosene, and to a certain extent, on charcoal as well as firewood. The high-income groups (VI and VII) depend mainly on LPG and electricity (Table 7.3).

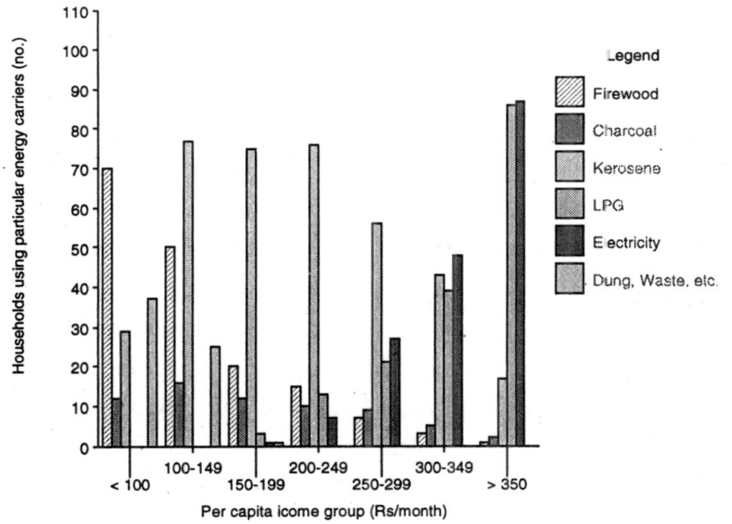

Figure 7.8: Households using various energy carriers.

Table 7.3: Utilisation indices of different income groups using different energy carriers.

Income group PCMI (Rs)		No of households in the range	Households using different fuels (no.)						
Group	Range		Firewood	Charcoal	Kerosene	LPG	Electricity	Dung,waste, etc	Total
I	<100	148	70 (0.47)	12 (0.08)	29 (0.2)	0	0	37 (0.25)	148(1.00)
II	100-149	168	50 (0.30)	16 (0.10)	77 (0.46)	0	0	25 (0.15)	168(1.00)
III	150-199	112	20 (0.18)	12 (0.11)	75 (0.67)	3(0.03)	1 (0.01)	1 (0.01)	112(1.00)
IV	200-249	121	15 (0.12)	10(0.08)	76(0.63)	13(0.11)	7 (0.06)	0	121(1.00)
V	250-299	120	7 (0.06)	9(0.08)	56(0.47)	21(0.18)	27(0.23)	0	120 (1.0)
VI	300-349	138	3 (0.02)	5(0.04)	43(0.31)	39(0.28)	48(0.35)	0	138 (1.0)
VII	>350	193	1 (0.01)	2(0.01)	17(0.09)	86(0.45)	87(0.45)	0	193(1.00)
Total		1000	166(0.17)	66(0.07)	373(0.37)	162(0.16)	170(0.17)	63(0.06)	1000(1.0)

Note: Numbers in parentheses represent utilisation indices.

Thus, the choices in energy consumption reveal a pattern of substitution of one carrier for another. With an increase in income, solid fuels (charcoal and firewood) give way to a liquid fuel (kerosene) which in turn is displaced by gas (LPG) and electricity which are the most desirable energy carriers for households (Figure 7.9). This may be because they are clean energy carriers and easy to operate. Also, substitution gradation by the households is found to be based on the efficiency of the device employed by the user for a particular carrier.

Energy Carrier Substitution

This analysis considers that firewood, charcoal, kerosene, LPG and electricity are five different energy carriers competing through their appropriate devices (stoves) for adoption by households. The likelihood of choosing a particular carrier depends on the economic and social status of the household and the decision here is discrete, i.e. selecting or not selecting. The problem, therefore, is of the discrete choice of the household for inter-fuel substitution where the household's characteristics play a major role. Thus, the analysis is used to study the fuel preferential characteristics of the households. A multilogit model has been used (see Chapter 3) to explain the probability of selection of the fuel for cooking and water-heating by the household.

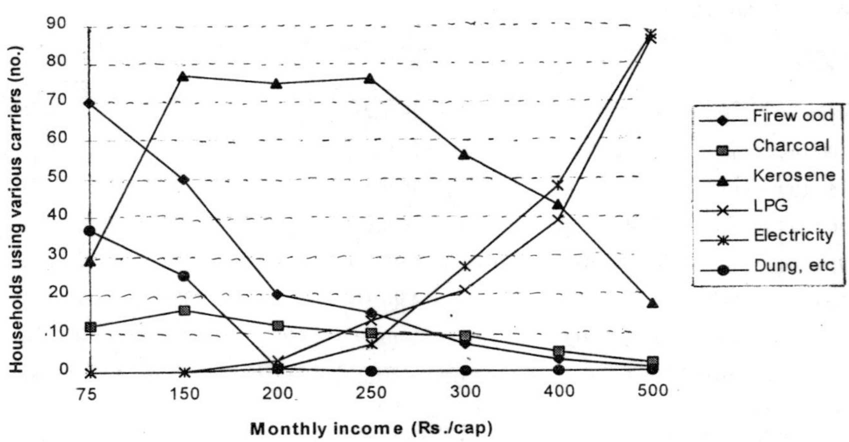

Figure 7.9: Substitution of energy carriers for cooking.

Table 7.4: Variables included in the model.

Symbol	Variable
Response variables	
P1	Firewood
P2	Charcoal
P3	Kerosene
P4	LPG
P5	Electricity
Independent variables	
FS	Size of the family
FI 1	Per capita income (dummy) for Rs < 150
FI 2	Per capita income (dummy) for Rs 150-299
FI 3	Per capita income (dummy) for Rs 300-500
FI 4	Per capita income (dummy) for > Rs 500
PR	Relative price of carrier (per energy unit) dummy if < 0.5 of other
AV	Availability dummy (1 if the energy carrier is available to the household
OCP	Occupation dummy (1 if the head of the household is employed
TR	Tradition dummy (1 if the household uses the energy carrier as a tradition

In this study, the dependent variable Y denotes the preference or otherwise of a particular fuel with respect to another fuel (for example, wood against LPG) and is thus a logical variable. The values, one and zero, can be assigned to Y depending on whether the fuel is preferred over the alternative fuel or not. The independent variable is the vector X denoting the characteristics of the household, i.e. size and income of the family, and other variables such as the price of the fuel (Table 7.4). It is important to note here that what we want to analyse is (i) not the choice of a particular energy carrier type over all the remaining carrier types, but (ii) identifying the factors that influence the choice of a particular carrier against a *specified alternative*. Therefore, we do not use a multinomial logit model which is more relevant for (i) rather, we use a series of simple logits to determine the choice between each pair of energy carriers.

This is a system of n non-linear equations where j is the number of explanatory variables. These equations were solved using the statistical subroutine package CO5NBF. The log-odds ratios of different fuel combinations are given in Table 7.5. It should be noted that the parameter estimates presented in the table indicate the effects of change in independent variables on the log-odds ratios, but not on the probabilities themselves. A non-zero estimate means that the percentage change in p_i and p_1 are not the same. For any p_i, however, a zero estimate does not mean that the probabilities do not change. In particular, the effect on a particular p_i depends on the magnitude and sign effects on all other log-odds ratio.

From the results presented in Table 7.5, it appears that a majority of the parameters are significant. From the table, it is discerned that kerosene is the most preferred fuel for cooking. Its choice over firewood is influenced by the income of the household and price of the fuel. The preference for firewood is influenced mostly by its availability, size of the family and occupation of the head of the household. The choice between kerosene and LPG is insensitive to family size and availability and income seems to be the most determinant factor (since the differential cost of the devices is very high).

Table 7.6 summarises the main results of the logistic estimation in terms of each parameter on the log-odds ratios involved. The tables present the sign of effect (positive, negative or zero) for only those factors that are found to have a statistically significant influence.

As anticipated, the effects of income have the same signs as expected. On the basis of the energy ladder hypothesis, each move from a lower to a higher income category leads to a greater probability of selecting kerosene over firewood, and a move into the highest income category increases the probability of choosing LPG or electricity over kerosene. It presupposes that the lower income groups will not tend to use them. As seen from Table 7.6, at the lower levels of income, there is a negative tendency for the household to use LPG or electricity. Family size and occupation of the head of the family also play a crucial role in the selection of a fuel. With an increase in the size of the household, there is a greater probability of firewood being preferred to kerosene in low- and middle-income groups.

Table 7.5: Regression results.

Explanatory variables	Dependent variables (natural log of odds ratio)								
	CH:FW	**KR:FW**	**LPG:FW**	**EL:FW**	**KR:CH**	**LPG:CH**	**EL:CH**	**LPG:KR**	**EL:KR**
FS	0.1 (0.05)	-1.4 (0.12)	-1.5 (1.03)	0.6 (2.42)	0.2 (1.32)	1.2 (1.34)	1.4 (2.52)	0.3 (1.14)	2.3 (0.12)
FI 1	-1.8 (1.21)	0.4 (2.08)	-11.4 (8.49)	-22.0 (8.6)	0.8 (1.2)	0.2 (0.9)	-0.9 (1.1)	0.2 (1.4)	0.7 (1.3)
FI 2	-0.6 (2.34)	1.7 (1.12)	0.4 (0.58)	3.2 (3.0)	1.7 (1.3)	0.2 (0.7)	1.3 (5.6)	-4.0 (3.4)	-0.8 (0.4)
FI 3	-14.1 (8.45)	12.4 (5.40)	3.7 (1.89)	-0.4 (0.3)	5.6 (2.4)	13.7 (3.5)	9.3 (3.8)	16.4 (5.6)	1.7 (2.3)
FI 4	1.2 (5.98)	12.4 (7.96)	4.2 (3.14)	2.4 (1.2)	1.8 (2.6)	25.2 (7.6)	10.1 (4.8)	9.9 (2.1)	2.0 (0.9)
PR	0.6 (1.24)	-1.6 (1.40)	1.1 (1.65)	0.6 (2.3)	-0.3 (1.2)	-1.2 (0.8)	-0.7 (1.8)	-0.9 (1.2)	-2.2 (6.4)
AV	-1.3 (7.96)	0.6 (2.43)	0.4 (0.98)	1.2 (2.3)	0.1 (1.4)	0.04 (0.8)	-3.2 (8.5)	6.7 (3.4)	-1.4 (1.6)
OCP	-3.3 (4.2)	-4.3 (1.3)	-0.8 (0.1)	2.4 (2.1)	0.1 (0.8)	0.05 (1.2)	-3.4 (1.3)	1.3 (1.1)	-2.3 (1.6)
TR	1.9 (0.85)	-3.4 (5.46)	-10.2 (4.5)	-0.1 (2.2)	-1.3 (0.9)	-1.2 (0.2)	-3.2 (3.4)	0.01 (1.3)	-8.2 (3.4)

Note: Figures in parentheses represent standard errors.

Table 7.6: Sign of parameter estimates.

Explanatory variables	Dependent variables (natural log of odds ratio)								
	CH:FW	**KR:FW**	**LPG:FW**	**EL:FW**	**KR:CH**	**LPG:CH**	**EL:CH**	**LPG:KR**	**EL:KR**
FS	-	-	-	0	0	0	0	0	+
FI 1	-	0	-	-	0	0	0	0	0
FI 2	0	+	0	-	+	0	0	-	-
FI 3	-	+	+	+	+	+	+	+	0
FI 4	0	+	+	+	0	+	+	+	+
PR	0	-	0	0	0	-	0	0	0
AV	0	0	0	0	0	0	0	+	0
OCP	0	-	-	+	0	-	0	+	-
TR	0	0	-	0	-	-	0	0	-

Note: + = significant positive effect at 10 per cent level (t > 1.282)
- = significant negative effect at 10 per cent level (t < -1.282), 0 = no significant effect

Energy Carrier Competition

The results of the multilogit model indicate that with increasing incomes, households switch over from one carrier and its associated device to another carrier and its associated device. Thus, the substitution of one carrier by another can be looked upon as a competition between two carrier-using devices for adoption by households so that the ratio of the number of households using the two devices is a function of the household income. While income is the main criterion, the technological (i.e. usage, convenience, etc.) superiority of the preferred carrier is also an important factor to be considered. Since the efficient technology usually costs more, or at least in first cost, the income criterion comes into play. With this perspective, another model for carrier substitution can now be developed to provide a greater insight into the substitution process.

In general, most technological changes can be described as a substitution of one material or energy carrier for another. In the case of household energy carriers, the changes boil down to creating new and better ways of satisfying the energy needs of households [1]. These needs tend to vary with changes in the monetary incomes of households. Thus, the competition between the old and new carriers may be viewed as a sequential process in which the old energy carriers give way to new ones.

A thorough understanding of the basic mechanism involved in the diffusion of new technologies is a prerequisite to any reliable assessment of the possible impact of technological developments upon the society in general. Also, long-term energy planning demands that we should have some idea of how any technological competition is likely to proceed, through which one can foresee its impact on the demand for different energy carriers. In forecasting the technological competition, and thereby the most likely events of the future, the curve-fitting technique may not take into account several important factors that affect the economic and other decision-making processes. To overcome this, a competition model is developed to evaluate the factors quantitatively. The model is described in Chapter 3 and its application is illustrated here in the case of competition of various energy carrier pairs.

The model of energy carrier competition in the residential sector has been attempted not only with a view to understanding the mechanism that governs the substitution process but also to study the factors that influence it. Even though it is generally accepted that logistic equations follow the Lotka-Volterra equations, it is an accepted fact that energy carrier competition is intimately related to economic considerations of households. With this in view, a competition model was developed to get a greater insight into the substitution process for Bangalore. This approach should provide a better understanding of the dynamics of the process of technological competition in the residential sector.

The proposed model describes the process of competition between different energy carriers competing through their appropriate cooking devices (stoves) for adoption by households. Such a multi-level process can be called the process of replacement of an evolutionary type.

The Lotka-Volterra equations describing the competing biological systems (ex:predator - prey system) are used in the present study to model the competing energy technologies (ex: firewood, kerosene, electricity, etc.) through their respective devices.

In the present case, we wish to study the competition of technologies in terms of the number of households associated with each of the energy carriers, old and new. Here, the variation is with income I rather than with time t. In the absence of competition, i.e. when the systems are isolated, we expect a gradual saturation. Thus, we arrive at a non-linear variation of the rate of changes with the actual population. The interaction with other technologies is assumed to sustain with increasing income. So, the technologies must oppose each other, but aid themselves, except for an inherent decay factor. Thus, all the concepts of technological competing systems have analogies in the biological systems, to a great extent.

When a new technology competes and replaces either fully or partially the old technology, the new technology is never an exact equivalent of the old one. Hence, it is not proper to compare the economics of these two through the unit prices. For example, in the case of competition between firewood and kerosene, it is not valid to compare the average price of a kerosene stove with that of a wood stove. In the first place, a firewood stove has no initial cost, transportation cost, etc. However, the efficiency, unbreakability and smokelessness of a kerosene stove influence the customer to opt for it. Thus, the alternative technologies do not compete on the basis of prices alone, but rather on the basis of maximum utility provided per rupee of cost.

The present model is compared with the observed data obtained from the household survey. It is seen that a second degree polynomial gives the best fit. The comparison of these best-fit curves with the empirical curves is shown in Figure 7.10 for two pairs of technologies (devices/stoves). It is clear from these figures that the theoretical prediction of a quadratic variation of ln (X_2/X_1) with income I is in agreement with the observations.

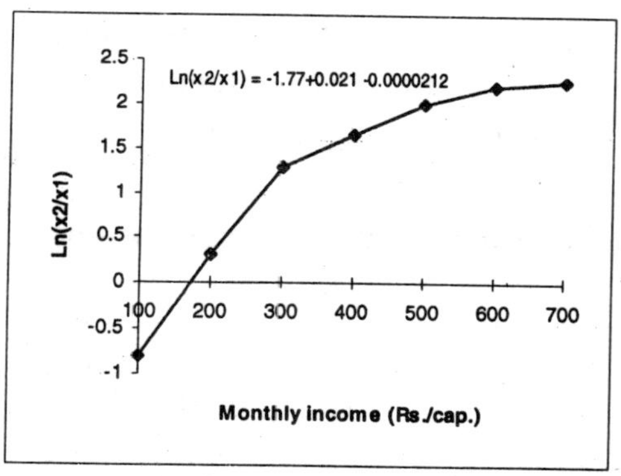

Kerosene for firewood.

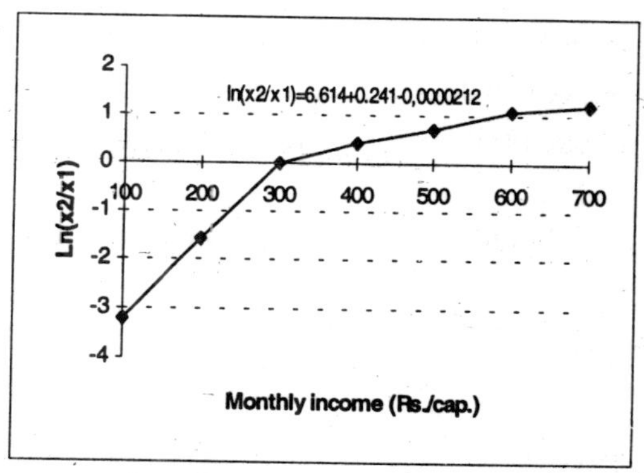

LPG for Kerosene.

Figure 7.10: Substitution of one carrier for another.

The following discussion explores the economic factors underlying the shifting of households from one energy-utilizing device to another.

Consumer Discount Rates and Energy Carrier Choices

Consumer discount rates are good indicators of the potential of energy conservation programmes. They can also be used to evaluate the penetration of different energy carriers and are useful in studying the carrier shifts. Here, the trade-off between the capital cost of the energy efficient device and the operating costs of the standard one is analysed using the household energy consumption data.

The choice of an energy carrier by a household depends mostly on the capital cost of the device (stove). Low-income households use firewood, charcoal, agricultural wastes, etc., whereas the high-income households use liquefied petroleum gas (LPG) and electricity (Reddy, 1990). The efficiencies with which these carriers are utilised (efficiency of the device) vary so also do the capital costs. Since the capital cost of efficient devices is high, low-income households cannot afford these, even if the return on investment is high. Thus, we can say that there is a relationship between the household income and the consumer discount rate.

Table 7.7 gives the performance characteristics of these carriers and their utilising devices. As the data show, a kerosene stove is more efficient than firewood stove but less efficient than an LPG stove. Thus, the useful energy consumption in the residential sector varies when we take the efficiency of a device into consideration.

Table 7.7: Energy consumption for cooking and water-heating through various devices.

Energy carrier	Final energy for cooking (TJ)	% of total	device efficiency (%)	Useful energy for cooking (TJ)	% of total	Final energy for water-heating (TJ)	% of total	Device efficiency (%)	Useful energy (TJ)	% of total
Firewood	1908	22.5	10	191	5.9	4255	63.9	10	425.5	24.7
Charcoal	499	5.9	18	90	2.8	315	4.7	18	56.7	3.3
Kerosene	4551	53.6	45	2048	63.1	1024	15.4	45	460.8	26.7
LPG	923	10.9	60	554	17.1	136	2.0	60	81.6	4.7
Electricity	609	7.2	60	365	11.3	934	14.0	75	700.5	40.6
Total	8490	100		3248	100	6664	100		1725	100

The capital cost of the device increases with increasing efficiency whereas the operating costs are more for the less efficient devices. Hence an initial large investment in the form of an efficient device would result in future savings. By determining how some of the carriers (in terms of efficient devices) must pay back the incremental investment, one can calculate the internal rate of return (IRR). The shorter the payback period, the larger the IRR.

In countries like India, households do not know their discount rates due to lack of knowledge regarding energy carriers and their utilising devices (Ruderman *et al.*, 1987). Thus, consumer discount rates are not easy to come by and, therefore, a significant contribution from the present study is the extraction of discount rates from the survey data. Using the life-cycle cost analysis, and the return on investment principles, the discount rates were obtained for the households. This is being done under the assumption that the consumer shifts from one carrier to another if his internal rate of return is greater than the consumer discount rate. Using the discount rates, the level of conservation equipment can be predicted.

The analysis assumes that consumer behaviour is characterised by the capital cost of the energy-utilising device. Even if the consumer does not decide on the basis of Life Cycle Cost (LCC), the discount rates can be used as a measure of market imperfection. This analysis examines different energy carriers (in terms of devices) used by different households and calculates the LCC taking into consideration the capital cost of the device and its life, operating cost, operating efficiency, energy carrier price, etc. In calculating the LCC it was assumed that the consumers do not anticipate any escalation in the prices of energy carriers.

By determining how some of the carriers (in terms of efficient devices) must pay back the incremental investment, one can calculate the internal rate of return (IRR). The shorter the payback period, the larger the IRR. This IRR is based on the assumption that the consumer is offered several investment options and he prefers the one with the lowest investment (Hausman, 1979). Since the LCC includes capital cost, operation and maintenance costs, it is easy to compare the cost of appliances whose performance characteristics are similar.

Though the value of IRR ($D_1 \rightarrow D_2$) reveals the benefits to a household by shifting from the device D_1 to the device D_2, it does not indicate whether the household decides that this internal rate of return is large enough to undertake the shift. Such decisions are based upon the consumer discount rate (CDR) used by the household and, in particular, upon how the IRR ($D_1 \rightarrow D_2$) compares with its CDR. The decision rule is as follows: if IRR ($D_1 \rightarrow D_2$) < CDR, then the switch $D_1 \rightarrow D_2$ is rejected, but on the other hand, if IRR ($D_1 \rightarrow D_2$) > CDR, then the switch is favoured.

A crude decision rule which does not take the time value of money into consideration is the simple payback period (SPP). The SPP is the number of years required to recoup the initial investment *I* through the annual savings *E*. Table 7.8 shows the SPP for different energy devices. It is seen that the SPPs are less than the 3–5 years usually considered to warrant investment. The SPP may indicate that a shift to energy-efficient devices is attractive, but it does not reveal the capacity of the household to make the initial investment. For this purpose, one must turn back to consumer discount rates.

Estimation of consumer discount rates

The consumer discount rate of a household depends upon its capital resources which determine whether the household can afford to initially invest an amount of capital (ΔI_0) on a more energy-efficient device in order to reap regular benefits later in the form of the energy savings (Dubin and McFadden, 1984). If this ability to defer current expenditures depends, in general, upon its income and, in particular, upon its disposable income, then it is obvious that the consumer discount rate must depend upon the income. If the income is high, then the household has the ability to make current investments in order to derive future benefits and, therefore, the consumer discount rate is small.

The data show that households shift from one energy carrier to another if their

Table 7.8: Simple payback periods through various device shifts.

Energy carrier	Carrier shift	Incremental cost of the device (Rs)	Savings in consumption (Gj/year)	Savings (*) (Rs)	Simple payback period (years)
Cooking					
Firewood	FW - KR	65	8.14	305.38	0.21
	FW - LPG	1090	10.76	403.46	2.70
	FW - EL	490	14.38	539.10	0.91
Charcoal	CH - LPG	1070	6.30	475.47	2.60
	CH - EL	470	9.91	746.67	0.78
Kerosene	KR - LPG	925	2.62	224.17	4.13
	KR - EL	425	6.23	534.19	0.80
Water-heating					
	FW-EL	588	8	283	0.48
	CH-EL	560	11	613	1.10
	KR-EL	520	4	163	0.31

Note: (*) Savings in 1990 prices.

income increases and the consumer discount rates decrease exponentially with household income. To test whether this assumption is in accordance with the observations, the χ^2 test was conducted and the result confirmed that the fit is significant at a 5 per cent confidence level (Table 7.9).

The following equation shows the relationship between consumer discount and income:

$$\ln (CDR) = 6.3822 \; -0.0082\, I$$
$$(9.37)$$

$R^2 = 0.92$

(Figures in parentheses represent *t* value)

Another way of testing the relationship is to compare the observed values with the empirical values of the consumer discount rates as a function of the household income. Figure 7.8 shows the number of households adopting various carriers/devices *versus* income. At the point where the curve for one carrier, say, firewood, intersects the curve for another carrier, say, kerosene, it is obvious that the number of households using firewood is exactly equal to the number of households using kerosene and, therefore, that the IRR (firewood --> kerosene) corresponding to that device switch defines the consumer discount rate at the income corresponding to the intersection of the firewood and kerosene curves. By the same logic, the consumer discount rates corresponding to other incomes can be obtained from the various intersection points in Figure 7.8. When the logarithms of the empirical consumer discount rates are plotted against the incomes, it is seen from a least-square fit that the relationship is indeed exponential as suggested above. (Figure 7.11)

Table 7.9: Consumer discount rates.

Income (Rs/cap/month)	Consumer discount rate
150	208
220	90
285	40
300	54
350	24
390	27
510	11

Energy Services

Residential energy services

Cooking accounts for the largest share of residential energy consumption. It accounts for about 10 per cent of the family's budget in the low-income groups and about 3 to 5 per cent of the higher income groups. Among the major carriers, kerosene was the major source of energy with 54 per cent of the total consumption followed by firewood with 22.5 per cent. Charcoal, LPG and electricity shared the remaining quantity. But firewood and charcoal are being used with low efficiencies of only about 10 and 18 per cent respectively[1]. Therefore, if the useful energy is considered, kerosene's share was 63 per cent and that of firewood only 6 per cent. The rest of the energy was from charcoal, LPG and electricity. (Figure 7.12)

Water-heating accounts for 6 to 8 per cent of a family's budget in the low-income groups and about 2 to 3 per cent in the high-income groups. Firewood and electricity are

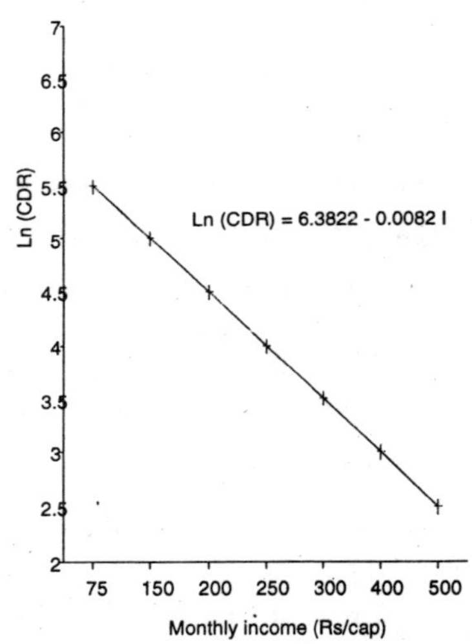

Figure 7.11: Dependence of consumer discount rate on household income

[1] This would imply that other advantages including competitive prices overcome the handicap of low efficiencies. Presumably these are identified and quantified to produce suggestions to modify the usage patterns in the larger societal or economic interest.

the main energy carriers used. In the final energy used for water-heating, 64 per cent was from firewood, 15.4 per cent from kerosene and 14 per cent from electricity. The case is reversed if the useful energy is considered, electricity sharing 40 per cent and firewood 25 per cent. (Figure 7.13)

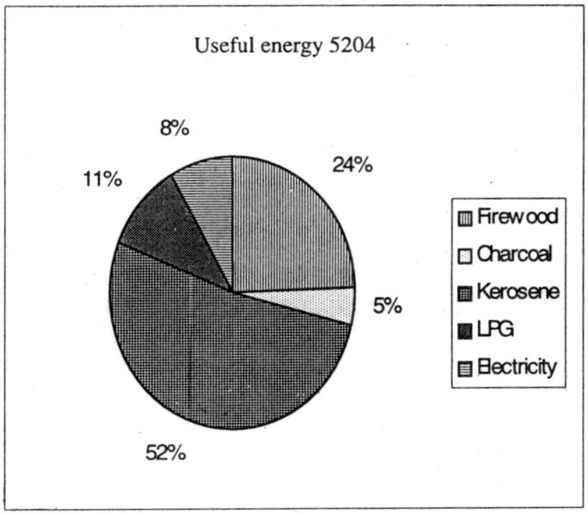

Cooking energy (1990)

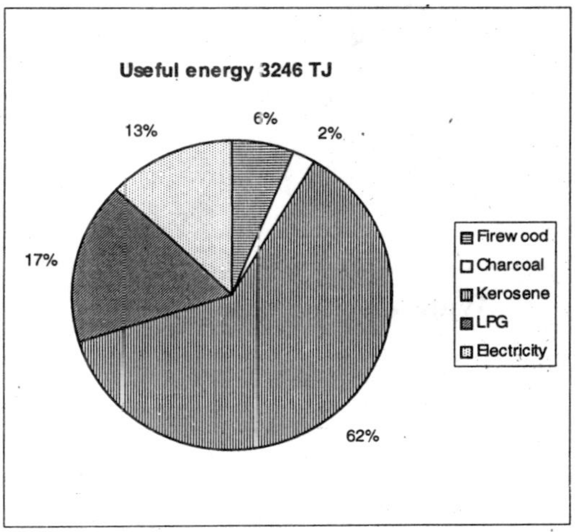

Cooking energy (1980)

Figure 7.12: Variation in cooking energy consumption.

Lighting

In the energy budget of a family, lighting accounts for 1 to 2 per cent. Kerosene and electricity are the only carriers used for lighting, the former being used by the low-income groups to the extent of 90 per cent while the other groups depend mainly on electricity. The percentage share of energy in the total consumption varies considerably, electricity accounting for 96 per cent.

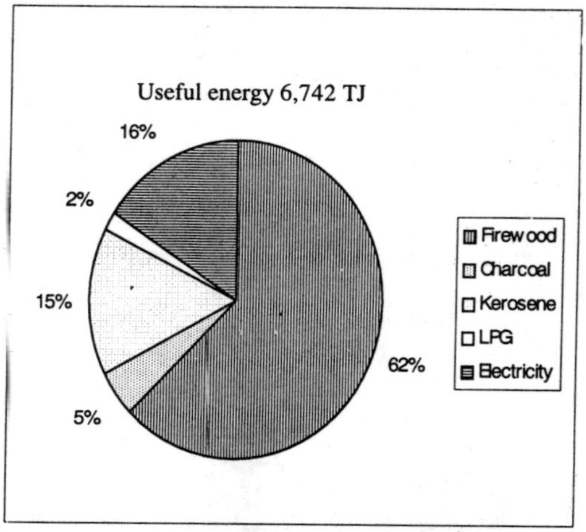

Water-heating energy (1990)

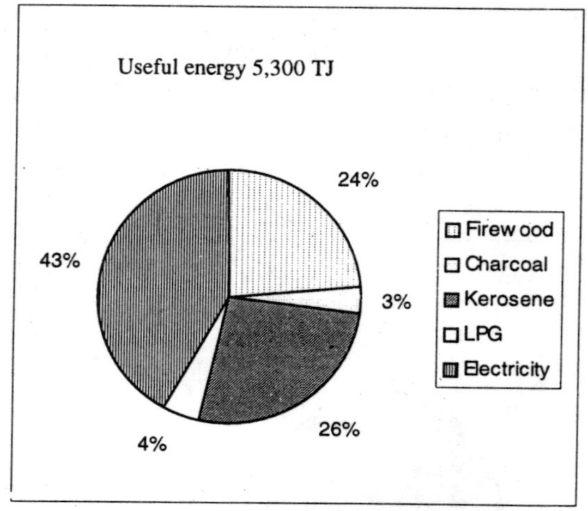

Water-heating energy (1980)

Figure 7.13: Variation in water heating energy.

Figure 7.14 shows a source-service matrix for the residential sector of Bangalore. In all, this sector consumed 13,107 TJ of final energy (1983-84) with contributions from firewood (47.3 per cent), kerosene (25.4 per cent) and electricity (12.9 per cent). If the useful energy is considered, then, kerosene's share was 40.5 per cent followed by electricity with 27.3 per cent while the share of firewood was a mere 16.8 per cent.

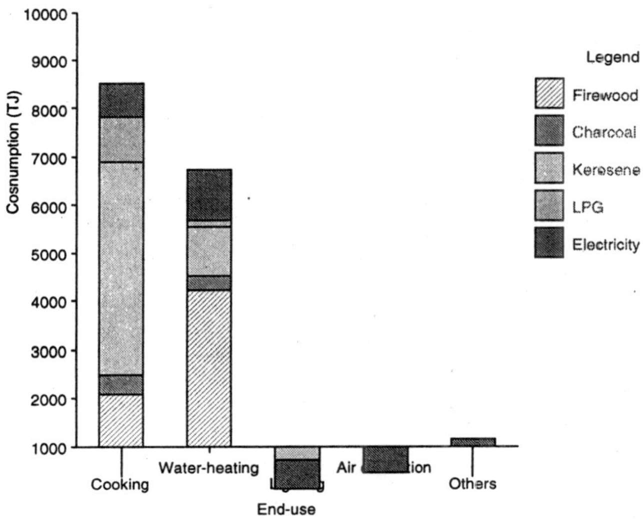

Figure 7.14: Energy source service .

Commercial Energy Services

The commercial sector includes such diverse institutions as hotels, office buildings, hospitals, theaters, etc. The energy carriers that are used in these institutions are firewood, charcoal, kerosene, LPG and electricity for end-uses like cooking, water-heating, lighting, space-cooling, etc. The total energy consumed in this sector was 1,578 TJ (1989-90). Of all the energy carriers, firewood headed the list with 741 TJ (46 per cent of the total) followed by charcoal with 260 TJ (16.5 per cent), electricity accounting for 16 per cent of the total. The contribution of different end-uses in the commercial sector was cooking 1,284 TJ (81.3 per cent), and water-heating, 72 TJ (4.6 per cent). The remaining quantity was shared by lighting, air-circulation, etc. (Figure 7.15).

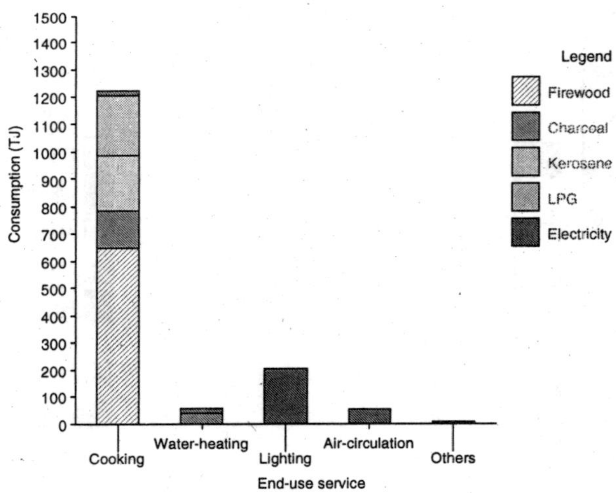

Figure 7.15: End-use analysis in commercial sector.

Industrial Energy Services

The energy carriers used in the industrial sector consists of firewood, charcoal, kerosene, LPG, electricity, coal, and petroleum products. The services provided by these energy carriers are process heat, lighting, space-cooling, compressing, pumping, etc. Of all the energy carriers, electricity and petroleum products with 3,565, and 3,565 TJ of consumption are the main energy carriers with a combined share of 72 per cent. In the end-uses, process heat heads the list with 80 per cent of the total while the remaining amount is accounted for by water-heating, lighting, blowing, shaftpower, etc (Figure 7.16).

Transport Energy Services

Transportation is an important function in an urban society. Of the total energy consumption, transportation accounts for 27 per cent. In total, 32 per cent of the transportation energy is used for private transport, 28 per cent for public transport and 40 per cent for goods transport. Petrol, diesel and coal are the only fuels used by automobiles and locomotives. In the total usage, petrol and diesel account for 88 per cent of the total consumption.

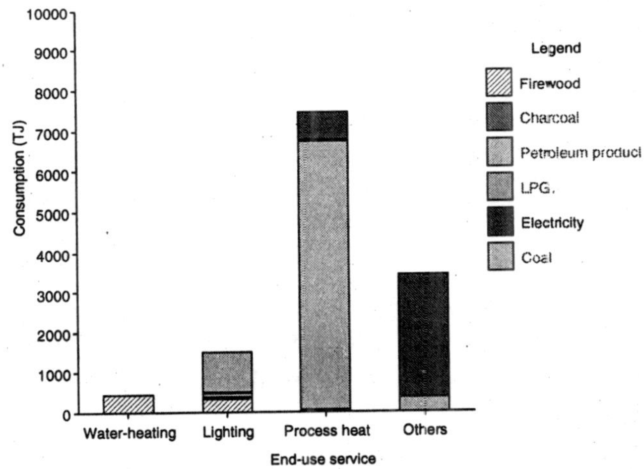

Figure 7.16: End-use analysis in the industrial sector.

Energy Source-service Matrix

Table 7.10 provides the energy **source-service matrix** (final energy in TJ) for Bangalore for the year 1989-90. The table also shows the quantity of each energy carrier utilised to provide energy service. Of all the carriers, petroleum products and electricity constitute nearly three-fourths of the total consumption. For example,. 1,390 TJ of electricity was used for lighting, 760 TJ for water-heating, 481 TJ for cooking, 775 TJ for process heating, etc. In all, 8,364 TJ of electricity was utilised to provide energy services to Bangalore. In the case of end-use consumption, cooking, water-heating, industrial process-heating and transport accounted for 80 per cent of the total. For example, for cooking, 9,617 TJ of energy was used to during 1989-90. Of this, 2,584 TJ was provided by firewood, 727 TJ by charcoal, 4,657 TJ by kerosene, 1,167 TJ by LPG and 609 TJ by electricity.

Energy Flow in Bangalore

Efficiencies of energy flow in Bangalore

In Bangalore, the major energy carriers that are being utilised are petroleum products (petrol, diesel, fuel oil, light diesel oil, LPG), electricity and firewood. They are being transported or transmitted from the source to the end-use point. During this flow three

Table 7.10: Energy source-service matrix for Bangalore (1989-90).

Energy carrier	End-use service						
	Cooking	Water-heating	Lighting	Process heat	Transport	Others	Total
Firewood	2584	4856		421			7861
Charcoal	727	379		70			1204
Petroleum products	4657	1094	224	6639	11491	29	24603
LPG	1167	145		108			1420
Electricity	481	760	1390	775		457	8364
Coal	0	0		1260	528	498	1788
Total (GJ)	9617(21.3)	7234(16.0)	1614(3.6)	9273(20.5)	12019(26.6)	5484(12.1)	45240(100)

Note: The numbers in parentheses represent percentages.

types of efficiencies are encountered (see Chapter 3), viz. simple energy (SEE), complex energy (SEE) and device efficiencies (DEE). These efficiencies provide information about the quantity of energy lost during the energy flow – from source to destination. The multi-energy input-out system is reduced to a and the calculations of energy flow are shown in Figure 7.17.

The CEE is always less than the corresponding SEE for all the resources. While kerosene and LPG are used with high device efficiencies, firewood and charcoal have been used with low efficiencies. It is to be noted that while the CEE of many of the resources is only marginally smaller, that of the grid electricity is substantially lower leading to a reversal in the rank ordering of the resources. An important finding is that the efficiency of using grid electricity can be much lower when additional energy inputs in producing and transmitting electricity are considered. The CEE of kerosene is quite low compared to its SEE. This is because substantial quantities of diesel are being employed for transporting kerosene from the source to the end-use point.

During 1989-90, for instance, Bangalore consumed 45,240 TJ of energy out of which 24,603 (54.5 per cent) came from petroleum products, 8065 TJ (20.1 per cent) from firewood and charcoal, 8,364 TJ (18.5 per cent) through electricity and 1,788 TJ (4.0 per cent) from coal. These energy carriers are transported or transmitted through trucks (26.8 per cent of total), railways (54.3 per cent) and through transmission lines (18.5 per cent). The transported energy is being distributed through the Government of

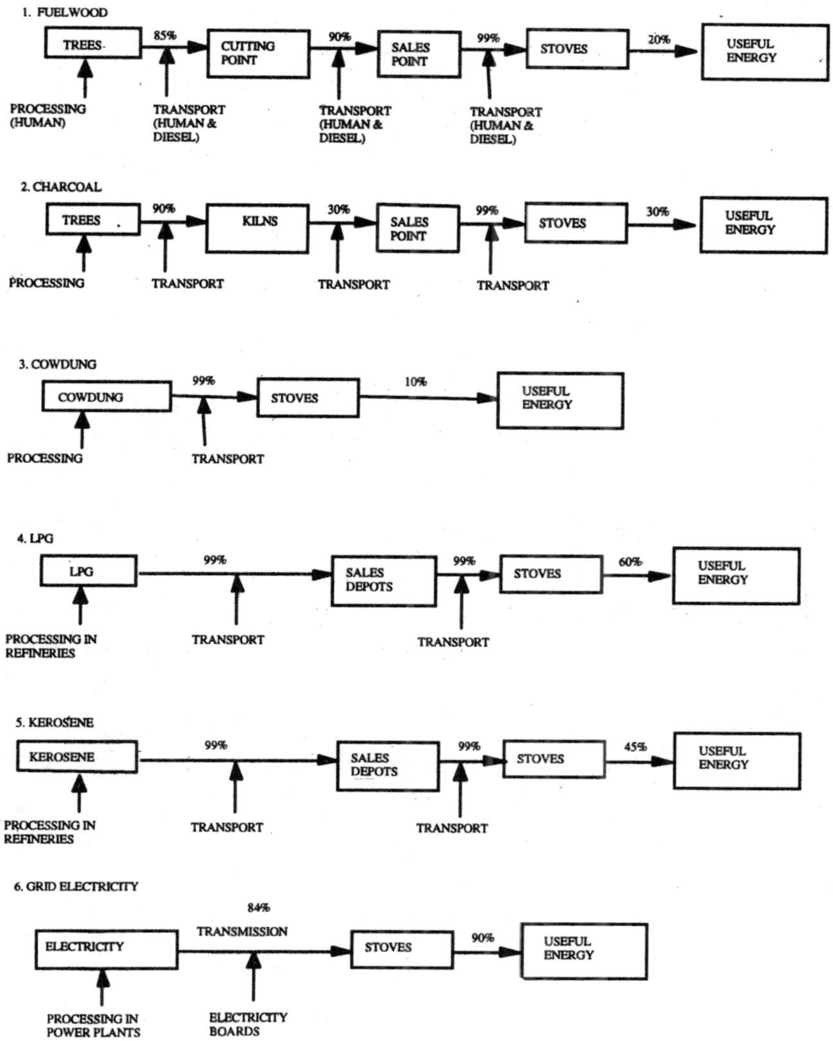

Figure 7.17: Efficiencies of energy flow.

India agencies (60.4 per cent), private contractors (18.9 per cent) and the Karnataka State Electricity Board (18.5 per cent). This energy is utilised in residential (38.1 per cent), industrial (29.3 per cent), transport (27.1 per cent) and commercial (3.5 per cent) sectors. Out of a total 45,240 TJ of energy, 21.2 per cent is used for cooking, 16 per cent for water-heating, 29 per cent for industrial applications, 27 per cent for transport and 3.6 per cent for lighting. However, all this energy is not utilised as useful energy. Most of it goes waste due to the inefficiencies of the utilising devices. Of the total energy, 59,625 TJ is converted into useful energy. Thus, the efficiency of energy flow is only 49 per cent[2]. (Figure 7.18)

Discussion

Implications of Bangalore's energy supply and consumption

Impact on forests

There is a difference between the usage of firewood and charcoal in urban and rural areas. Rural households depend on twigs and branches whereas the urban areas use logs which usually require the felling of trees. Thus, urban firewood and charcoal consumption has a much greater negative environmental impact as compared to the rural use.

Bangalore consumes around 1,200 tonnes of firewood per day which corresponds to deforestation of about 3,650 hectares or 10 hectares per day assuming that the growing stock per hectare in such a forest has a high value of 120 tonnes/ha (Reddy, A. K. N. and Reddy, B. S, 1983). In the case of charcoal consumption, the efficiency of conversion of oven-dry wood into charcoal is about 30 per cent in the type of kilns producing charcoal which is used in Bangalore. To supply 130 tonnes of charcoal per day about 470 tonnes of wood have to be carbonised every day for which 4 hectares of forest has to be cleared. This indicates that about 14 hectares of forest would have to be cleared each day to keep Bangalore supplied with its firewood and charcoal demand if the wood is being obtained non-renewably by clearing forests.

---

[2] This considers only device efficiency. If the efficiency of supply, transportation and distribution are considered, the system efficiency will be much less.

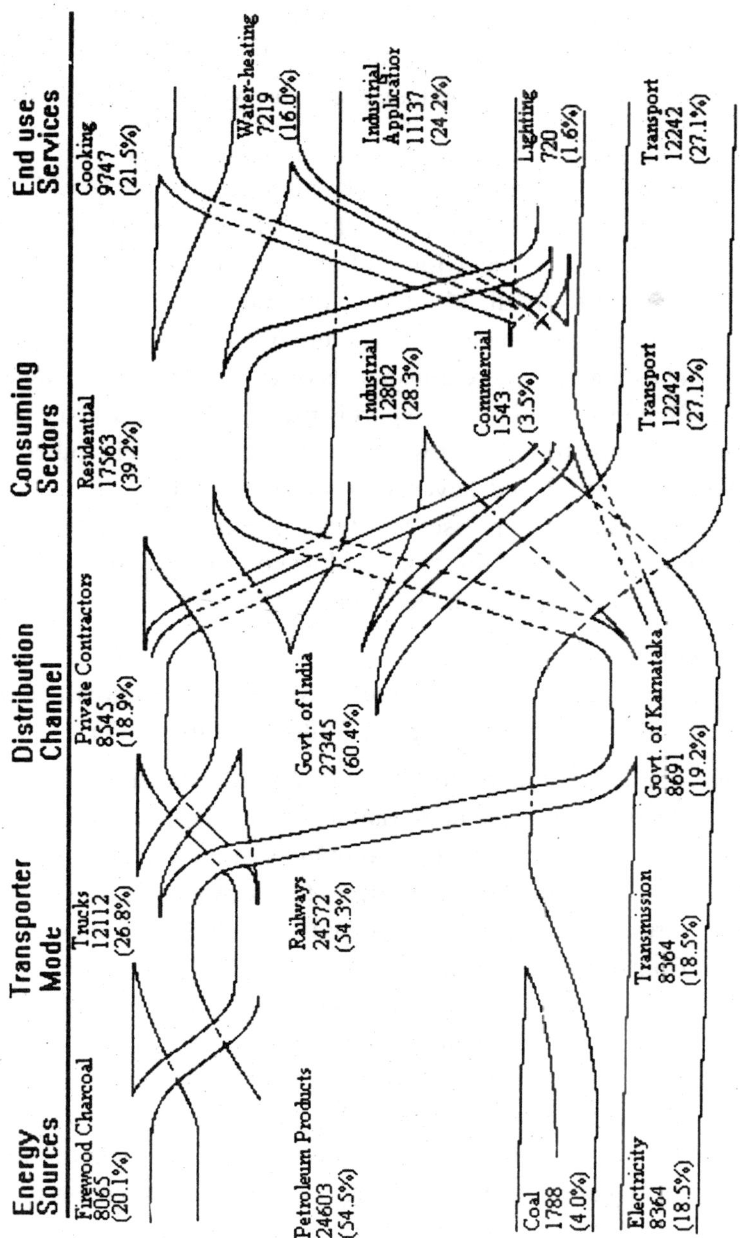

Figure 7.18: Flow of energy (1989-90).

Impact on the transport system

The transport of various fuels into Bangalore involves truck and wagon traffic. The records of weigh-bridges (for trucks) and railway goods sheds (for railway wagons) for the months of October-November, 1984, indicated that the total number of trucks and railway wagons bringing freight into Bangalore every day was 630 + 81 and 125 + 37 (Anon, 1984).

It follows that about 18 per cent of the truck and 8 per cent of the railway wagon traffic for firewood, 3 per cent of truck traffic for charcoal, 6 per cent of railway wagon traffic for petroleum products, 1 per cent for coal and 1.3 per cent of truck traffic for LPG are committed to transport. In all, 23.3 per cent of the trucks and 15.1 per cent of the railway wagons that bring into Bangalore its commodities are tied up for transporting energy carriers. Thus, in a situation where transport is often a constraint on the economy, a significant fraction of both road and rail transport is used for the supply of fuels. There is an *opportunity cost* with this transport. Instead of transporting fuels, other essentials like foodgrains, medicines, etc., could have been transported. Thus, the cost that the society is incurring by not transporting essential items is considered as the opportunity cost associated with the withholding of this truck and wagon capacity from the economy, but this opportunity cost has not been calculated here.

Impact on transport energy

To transport various energy carriers from source to destination requires energy in the form of diesel. These energy costs are estimated as follows: (Table 7.11)

On an average, about 1,200 tonnes/day of firewood was transported to Bangalore through trucks and railways from distances ranging from 35 to 700 km. All this constituted a total of about 156,772 tonne km/day which, at an average of 22.04 tonne km/kg diesel for truck and 154 tonne km/kg of diesel for railway wagons (*Encyclopedia of Energy*, 1971), requires 5.07 tonnes of diesel/day. The 132 tonnes of charcoal supplied to Bangalore from varying distances (ranging from 70 to 250 km) constituted about 26,404 tonne km for which 1.23 tonnes of diesel was used every day. Bangalore consumed about 1,168 tonnes of petroleum products per day which were brought from Madras situated 350 km away through railway wagons and this corresponds to about 412,340 tonne km. The total quantity of diesel used to transport petroleum products was 2.65 tonnes per day. The transport of 50 metric tonnes of LPG is done through trucks from Bombay and Koyali (where the LPG refineries are located) which are at distances of 1,000 and 1,200 km., respectively, with a diesel consumption of about 0.63 tonnes per day. Thus, the total energy consumption through trucks and railways for transporting energy carriers to Bangalore comes to about 11.5 tonnes of diesel per day.

Karnataka's electricity is obtained from different generating stations in the State and also, to some extent, from other States. Considering the transmission and distribution losses of 21.5 per cent (Anon, 1988), the generation of one kWh of electricity is reduced to 0.785 kWh where it is delivered to the consumers.

The energy content of the average daily consumption of 11.93 tonnes of diesel is 472 GJ on the basis of 38 GJ/ tonne. Also, 3,021 GJ is lost in transmitting the electricity. This expenditure of energy leads to the supply of 92,827 GJ. But not all of this energy which is supplied to the consumer is utilized for the desired end-uses, e.g. cooking. The thermal efficiency of stoves and other end-use devices varies from 10 to 70 per cent. Thus, the energy delivered is only 46,143 GJ per day in comparison with 3,493 per day of energy used to supply it. In other words, to deliver 100 units of energy about 7.6 units are being expended (Table 7.11).

Table 7.11: Energy costs of transporting various carriers (1989-90).

Energy carrier	Mode of transport	Quantity supplied	Distance (km)	Tonne-km	Consumption norm (tonne-km/kg)	Diesel consumption (tonnes)	per cent of total
Firewood	Truck	420	35	14700	22.04	0.67	
		600	135	81000		3.68	43.19
		24	350	8400		0.38	
	Total	1044		104100		4.72	
	Rail	56	212	11872	154	0.08	
		60	680	40800		0.26	
	Total	116		52672		0.34	
	Total (Truck+rail)	1160		156772		5.07	
Charcoal	Truck	48	220	10560	22.04	0.48	
		45	250	11250		0.51	10.50
		22	70	1540		0.07	
		21	180	3780		0.17	
	Total	136		27130		1.23	
Coal	Rail	130	1000	130000	154	0.84	
		65	1200	78000		0.51	16.91
		65	1500	97500		0.63	
	Total	260		305500		1.98	
LPG	Truck	50	350	17500	22.04	0.79	6.77
Petroleum products	Rail	1168	350	408800	154	2.65	22.63
Grand total				915702		11.73	100.00

Table 7.12: Transportation energy for various energy carriers.

Energy carrier	Unit	Quantity supplied	Quantity (GJ)	Device efficiency (%)	Useful energy (TJ)	Transport/ trans-mission energy (TJ)	Transport energy as a per cent of useful energy
Firewood	tonnes	1422	22752	10	2275.2	254	11.2
Charcoal	tonnes	120	3972	18	715	47	6.6
Coal	tonnes	200	7000	60	4200	66	1.6
Petroleum products	tonnes	1840	69920	40-80	41952	181	0.4
LPG	tonnes	50	2250	60	1350	58	4.3
Electricity	kWh		14050	60-70	9133	4934	54.0
Total			119944		59625	5541	9.3

Note: For electricity, transmission losses are considered.

Impact on foreign exchange

The 5.3 million litres of diesel/year utilized for transporting Bangalore's energy cost about Rs. 75.5 lakhs at an international rate of $14.4 per barrel of crude (Feb. 1988 price) or Rs.1.48/litre of diesel. Since about one-third of India's oil is imported and paid for in foreign exchange, it can be argued that Rs.25.2 lakhs or about $84,000/year was the foreign exchange expenditure on supplying Bangalore with its energy requirements.

Impact on environment

In Bangalore, transportation energy, particularly by road vehicles, constitutes nearly 30 per cent of the total energy consumption out of which 80 per cent is through passenger transport. Between 1980–90, there was a significant growth in personal transport, particularly two-wheelers (15.41 per cent p.a.) which accounted for 60 per cent of the total vehicle kilometres. This growth in personal transport, which has a high energy intensity, increases energy consumption. Thus, with an increasing use of transport services, the level of energy consumption, and thereby, the atmospheric pollution increases. Carbon monoxide, unburnt hydrocarbons, etc., emitted from vehicles contribute to atmospheric pollution. The transportation data of Bangalore shows that the contribution of two-wheelers to the unburnt hydrocarbons is very high (55.4 per cent of the total).

Implications of energy carrier substitution

The results of the study show that petroleum products and electricity are two important energy carriers used by an Indian metropolis, even though significant quantities of firewood and charcoal are being utilised. In the sectoral consumption, the residential sector contributes a major portion of the total.

The results of the energy carrier choice in the residential sector of Bangalore confirm the hypothesis of the energy ladder concept. Each step of the ladder corresponds to a different and more sophisticated energy carrier and the step to which the household climbs the ladder depends mainly on its income. The height of the step is determined by the economic factors like the initial cost of the fuel-utilising device, price of the energy carrier, and household energy consumption. In the residential sector, the choice of an energy carrier is discrete and it is conditional on the characteristics and environment of the household. The logistic model provides greater insights into the decision-making nature of the household.

The results of the competition model suggest that the substitution of energy carriers can at best be considered an economic phenomenon. The description is based on the premise that the substitution is the result of competition between various types of carrier-utilising devices where the efficient device takes over. A set of non-linear differential equations describe the variation of the number of households utilising various carriers with income and depend on the severity of the competition among them. These results fully describe the given substitution process corresponding to a particular set of parameters. This model appears to be very general and in conformity with the observed data.

The dynamics of the process of substitution of a new technology for the old one is important in a number of planning-related activities which include the resource allocation for research and development, technological forecasting and assessment, etc. The model developed here will provide a structure for predicting technological substitution. If such a technological substitution has been initiated, forecasts based on the relevant data indicate the rate at which the substitution will proceed in the future depending upon the economic conditions of the households.

Income-dependent technological competition is a characteristic of a stratified society. But as it becomes more egalitarian, this type of competition process disappears. Although competition of energy technologies can be envisaged as an evolutionary process, the government should provide significant subsidies or other financial assistance to eliminate the use of inefficient devices. Since efficient devices are expensive, reduction in capital costs through subsidies, rebates, etc., can induce poorer households to shift to more efficient and energy-conserving devices. This will also decrease the fuel-wood consumption thereby reducing the stress on forest resources.

Income-dependent consumer discount rates for the carrier shifts for cooking and water-heating imply that the cost aspect dominates the decision of Bangalore's households regarding the choice of an energy carrier. On the basis of this assumption, and given Bangalore's carrier prices, the theory predicts that the fuel shifts should be in a particular order, for example, firewood ---> kerosene ---> LPG. These shifts are consistent with the shifts empirically obtained from the household survey. Hence, as far as Bangalore's households are concerned, it is very reasonable to assert that their choice of energy carriers is determined by cost considerations.

However, the assumption that the cost aspect dominates Bangalore's households is not identical to the assumption that the users are completely indifferent to different carriers, except as regards costs (Kenneth Train, 1985). This is because, along with a decrease in the costs, there are also improvements in cooking convenience associated with a shift from solid fuels (firewood and charcoal) to a liquid fuel (kerosene) to gas (LPG) and electricity (instant lighting or starting of the stove, ease of turning the power up (or down) for boiling (or simmering) and smokelessness and cleanliness. In Bangalore, where cost minimisation automatically ensures convenience maximisation, cost is a good enough proxy for these other preferable characteristics of energy carriers.

When the carrier prices change, the IRR (D_1 ---> D_2) will also change. For example, if the LPG price increases while the kerosene price remains constant, the IRR (kerosene ---> LPG) will decrease and, after a certain LPG price, will even become negative, i.e. the carrier shift will be in the opposite direction because IRR (LPG ---> kerosene) is positive (Figure 7.19).

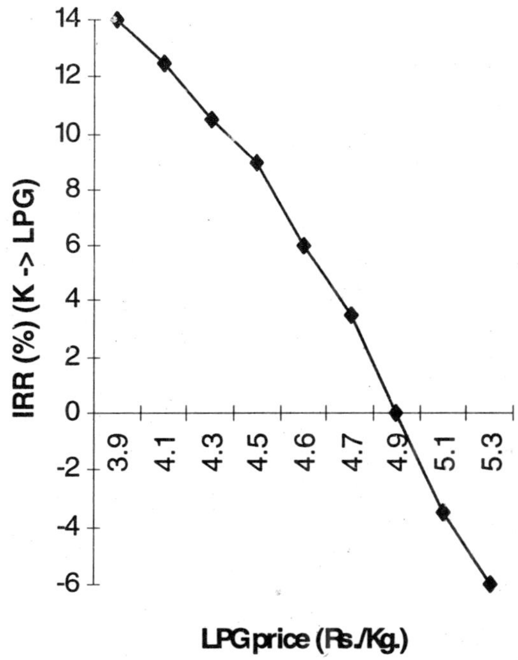

Figure 7.19: Effect of LPG price on IRR (K --->LPG).

In spite of the IRR (kerosene ---> LPG) becoming negative, if consumers still choose LPG over kerosene, even though their fuel costs increase when they switch over from kerosene to LPG, it obviously means that the consumer is prepared to pay a higher cost for greater convenience. Thus, according to the approach adopted here, such situations are conceivable given certain variations in prices. Under such conditions, it is clear that cost is not a proxy for the other preferable characteristics of energy carriers such as convenience and that cost minimisation does not automatically ensure convenience maximisation. Trade-off costs for convenience are unavoidable.

These results also imply that energy conservation through improved technologies may not succeed if consumers have high discount rates. Using the consumer discount rates, one can arrive at the proper incentives that are most cost-effective to induce households. The fact that emerges from this study is that in making decisions, which involve discounting over time, individuals behave in a manner which implies a much higher discount rate than can be explained in terms of the opportunity costs of the funds available in credit markets. Since these consumer discount rates substantially exceed the social discount rates used in benefit-cost calculations, the difference might be narrowed by policies which lead to the purchase of more energy efficient devices. Financial incentives like rebates, tax subsidies, etc., can lower the initial capital cost and make the trade-off toward a lower operating cost more favourable. Another possibility will be for the government or electricity boards to install energy-efficient equipment in households and collect the payments in monthly instalments. However, the consumer's knowledge regarding the costs, benefits, etc., also plays a significant role in the conservation programmes. The government, therefore, should try to educate the consumers in understanding the trade-off between the capital cost of the efficient device and the future energy savings.

8

THE POINTERS

The earth has enough for every man's need
but not for every man's greed.

– Mahatma Gandhi

Socio-economic policy decisions have not taken into account the long-term implications of urbanisation. It is a known fact that people migrate from rural to urban areas mainly for monetary gains. Thus, their decision to migrate can be considered as mostly economic and the efforts to make it worthwhile for them to stay in rural areas and depend on agriculture (through lower electricity tariffs and fertiliser subsidies) would be uneconomic. That it is being done nonetheless can be attributed to political pressures and other irrational factors. Large funds are generously made available to cities (as is evident by the mega-city projects in India). But sufficient thought does not seem to have been given to the costs of energy supply, changing land-use pattern, traffic congestions, pollution and ill health of the modern exploding cities. Planners consider these really serious but continue to believe that subsidised rural areas are grossly uneconomic.

It is not surprising, therefore, that, in the name of *economic rationality,* all urban areas are being turned into large untidy regions which are difficult to manage and control. They are deprived of energy supplies, constrained by mass transportation, choked by air pollution and troubled by religious and ethnic strife. And yet, the influx to urban centres continues unabated and the urban population is expected to double by the beginning of the next century! Why? The answer is obvious. It is uneconomic to resettle rural areas.

As a result, the most important conflict in the developing countries today is not between capital and labour, but between rural and urban regions. The rural areas contain most of the resources; the urban sector contains the power. So the urban classes have been able to grow at the cost of their rural counterparts. But in doing so, they have made the development process slow and unfair.

Fortunately, the view is gaining ground that cities with a large number of inhabitants may not be sustainable in the long run and unless resources are distributed optimally between rural and urban areas, the consequences would be disastrous for the mankind itself.

Energy in a Stratified Society – Lessons to be Learnt

The results of this study clearly illustrate the income-dependent nature of fuel utilisation by urban households – the rich use LPG and electricity whereas the poor depend on wood-based fuels. At present no policy instrument has been wielded to regulate the flow of biomass fuels, when the benefit of a large quantum of electrical energy is given to it. In fact, the supply of biomass fuels is almost left to the market forces. The major social cost of this private enterprise is environmental degradation through deforestation. This is a further illustration of a well-known phenomenon: economic inequalities result in negative environmental impacts at both ends of the income spectrum; the rich degrade the environment through wasteful and irrational use of resources, and the poor through the necessity of having to survive at the expense of the environment. In addition, the transport of wood-based fuels is associated with the consumption of a high-quality petroleum-based energy carrier, viz. diesel, and expenditure of foreign exchange on account of oil imports.

Thus, there are heavy social costs resulting from the poor having no alternative to firewood for cooking and water-heating. Even this firewood is indirectly subsidised in the form of diesel prices, and the cost of firewood often includes only that of extraction, transportation and distribution but not the cost of growing trees. Despite these hidden subsidies, the poorest sections have to spend as much as 17 per cent of their income on firewood. If the subsidies are removed as an isolated measure unaccompanied by other positive steps to provide fuel to the poor, it is obvious that their hardship will increase even more. Thus, a simplistic *remove-the-subsidies-and-get-the prices-right* approach often recommended by international funding agencies like the World Bank and International Monetary Fund (IMF) is likely to aggravate the sufferings of the poor.

The use of biomass fuels by poor households also illustrates another principle: energy consumption in an income-stratified society is far higher than would be the case in an egalitarian system[1]. This is because of the inefficiency with which energy is used by the poorest sections in contrast to the efficient energy carriers and devices used by the rich. In the case of cooking, for example, the efficiency of firewood stoves used by the poor is only about one-sixth that of the LPG stoves in the homes of the affluent. Thus, the over-all energy consumption for cooking would be much less if there are efficiency improvements in the energy carriers and devices used by the poor.

Energy Ladder and Carrier Substitution: What is in Store for Future?

The results of the energy carrier choice in the residential sector of Bangalore confirms the hypothesis of the energy ladder concept. Each step of the ladder corresponds to a different and more sophisticated energy carrier and the step to which the household climbs the ladder depends mainly on its income. The height of the step is determined by economic factors like the initial cost of the fuel-utilising device, price of the energy carrier and household energy consumption. The results of the competition model suggest that the substitution of energy carriers can be best considered as an economic phenomenon. The description is based on the premise that the substitution is the result of competition between various types of carrier-utilising devices where the efficient device takes over.

The dynamics of the process of substitution of a new technology for an old one is of importance in a number of planning-related activities which include resource allocation for research and development, technological forecasting and assessment, etc. The model developed here will provide a structure for predicting technological substitution. If such a technological substitution has been initiated, forecasts based on

---

1 Western industrialised societies are also economically stratified perhaps to a degree comparable to India. The main and crucial difference is that their lowest strata do not use the most rudimentary technologies.

the relevant data indicate the rate at which the substitution will proceed in the future depending upon the economic conditions of the households. Hopefully, as societies become more egalitarian this energy ladder concept based on income may disappear.

The income-dependent technological competition is very much a characteristic of a stratified society. Although competition of energy technologies can be envisaged as an evolutionary process, the government should provide significant subsidies or other financial assistance to eliminate the use of inefficient devices. Since efficient devices are expensive, reduction in capital costs through subsidies, rebates, etc., can induce the poorer households to shift to more efficient and energy-conserving devices. This will also decrease fuelwood consumption thereby reducing the stress on forest resources.

The income-dependent consumer discount rates for the carrier shifts for cooking and water-heating imply that the cost aspect dominates the decisions of households regarding the choice of an energy carrier. On the basis of this assumption, and given the carrier prices, the theory predicts that the fuel shifts should be in a particular order, for example, firewood ---> kerosene --> LPG. These shifts are consistent with the shifts obtained empirically from the household survey. Hence, it is very reasonable to assert that the choices of energy carriers of households are determined by cost considerations.

The assumption that the cost aspect dominates the households' choice is not identical to the assumption that the users are completely indifferent to different carriers, except as regards costs (Kenneth Train, 1985). This is because, along with the decrease in costs, there are also improvements in cooking convenience associated with the shift from solid fuels (firewood and charcoal) to a liquid fuel (kerosene) to gas (LPG) and electricity (instant lighting and starting of the stove, the ease of turning the power up (or down) for boiling (or simmering) and smokelessness and cleanliness. In urban areas, where cost minimisation automatically ensures convenience maximisation, cost is a good enough proxy for these other preferable characteristics of energy carriers.

Energy-Environment Nexus

The rapid growth of urban areas has led to the violation of ambient air quality standards and is mainly responsible for the degradation of the environment. According to UN study (Anon, 1989) morbidity rates are higher in the urban areas of the developing countries when compared to those of the developed countries. Also, the concentration of air pollutants exceeded the World Health Organisation's guidelines and the major source of air pollution was found to be the automobile.

Many urban areas are confronted with the emerging problem of air pollution. This is the result of several factors: growth in the vehicle population (for example, two-wheelers contribute one-third of the total CO, HC and NO_X emissions in Bangalore), the type of fuel used in them, the lack of strict control over vehicle maintenance, and the use of coal and furnace oil as fuel in industries. The failure of the public transport system to cope with increasing traffic levels is resulting in an unprecedented vehicle growth. While the developed countries are phasing out petrol and diesel vehicles, the developing countries are entirely dependent on them. Unless personal transport is reduced, the level of energy consumption and the resultant atmospheric pollution will continue to increase. Thus, heavy reliance on petrol in the developing world contributes to pollution-related health problems in large cities. The future, with the leaps in the urban population, and thereby the transport vehicles, is even less promising for the urban environment without proper policies to control and manage them.

Environmental degradation arises as an economic problem and may be attributed to the so-called *market failure*. In the present system, there is no penalty on a consumer for degrading the environment. He is largely ignorant about environmental factors and their implications on society. These social costs include health hazards, reduced output by workers, etc. The environment is considered by many as common property which cannot be priced. Thus, there is *competition to overuse* it.

An important implication is that environmental considerations should be treated as a part of the urban planning process. To reduce pollution levels, mass transport facilities must be improved significantly. Taxes should be imposed on those polluting the environment. This should be equal to that part of the cost (marginal social cost) which at present is not included in the production or import price. Energy subsidies should also be curtailed to highly polluting industries so that excessive energy-related emissions of pollutants are reduced. A system of emission tax along with an efficient use of energy and price reforms is a pragmatic approach to the control of urban air pollution. These are very sound policies as they would not only reduce pollution levels but also raise revenues. In the long run, it is better to reduce dependence on petroleum products and try for alternative technologies like compressed natural gas (CNG) and electric vehicles which are environment friendly. These approaches can act as effective and low-cost means of curbing this important source of air pollution.

Demand Side Management: Need for Long-Term Thinking

As the total energy consumption of urban agglomerations is growing at a rapid rate the immediate need is a conservation policy. From a long-term viewpoint, the first priority should be the efficient utilisation of energy resources. To achieve self-sufficiency and reduce energy consumption levels, the limited energy resources should be deployed more efficiently so that the physical quality of life and a higher development level can be achieved with a relatively small step-up in energy. To achieve this, the demand management must be directed towards the introduction of high-efficiency cooking and lighting devices. The government should induce industries to adopt efficient technologies which consume less energy per unit of output. The use of improved motors in industries, replacement of electricity for cooking and water-heating with other sources, etc., will reduce energy intensities significantly. Indeed, it is precisely to seek measures such as these that an end-use-orientation approach has been proposed (Reddy, *et al.*, 1991).

It is in this context that the appliance-wise disaggregation and appliance elasticities obtained in this study are significant. Without these, policy interventions are bound to lack an empirical foundation. For example, a water-heater has a high appliance elasticity and, in the total electricity demand, water-heating occupies the first place. And it would be a great mistake to continue using high-grade electricity for applications such as water-heating which could be achieved with low-grade fuels. Therefore, due importance should be given to the reduction of energy consumption in this service through substitution in the form of renewable energy sources like solar energy. For other services like lighting, the replacement of incandescent lamps with compact fluorescent lamps should be given prime importance. Once this substitution of energy carriers and replacement of inefficient appliances with efficient ones takes place, the appliance elasticities will decrease automatically. Thus, the increases in electricity consumption brought about by increases in the appliance stock can be reduced, or even prevented.

Even the poor want to switch over to carriers like LPG and electricity, but the initial costs of devices using these carriers is much higher than of firewood stoves and, therefore, the poor have no alternative to firewood. The only viable alternative is to assign the highest priority to satisfying the energy needs of the poor, in particular their need for inexpensive energy carriers and devices for cooking and water-heating. Fortunately, the technical means of achieving this objective are available; what is necessary is the will and the implementation machinery to help the poor to shift to alternative devices. Energy usage in an urban area is, therefore, one more illustration of the fact that when there are grave economic inequalities, the solution to the resulting problems consists in measures to satisfy the basic needs and remove these inequalities.

Role of Government and Institutions

The energy carrier substitutional model developed in this study provides insights into the decision-making nature of the household. It is from this perspective that the policy-makers should think in terms of increasing the pace of diffusion of energy-efficient fuel-wood stoves for those who cannot afford to purchase LPG or electric stoves. For this to happen, policy intervention must be directed towards altering the economic factors. This can be done by: (i) increasing the availability and reducing the prices of energy carriers like LPG and electricity and (ii) reducing the initial cost of LPG and electric stoves. In the long run, these measures would certainly decrease the demand for fuelwood at least in the urban areas. Thus, the real merit of the energy ladder concept is that it serves as a guide to policy formulation and intervention.

These results also imply that energy conservation through improved technologies may not succeed if consumers have high discount rates. Using the consumer discount rates, one can arrive at proper incentives which are most cost-effective to attract the households. The fact that emerges from this study is that in making decisions, which involve discounting over time, individuals behave in a manner that implies a much higher discount rate than can be explained in terms of the opportunity costs of the funds available in credit markets. Since these consumer discount rates substantially exceed the social discount rates used in benefit--cost calculations, the difference might be narrowed by policies which lead to the purchase of more energy-efficient devices. Financial incentives like rebates, tax subsidies, etc., can lower the initial capital cost and make the trade-off toward a lower operating cost more favourable. Another possibility will be for the government or electricity boards to install energy-efficient equipment in households and collect the payments in monthly installments. Also, the consumer's knowledge about costs, benefits, etc., also plays a significant role in the conservation programmes. The government should, therefore, try to educate the consumers in understanding the trade-off between the capital cost of the efficient device and the future energy savings.

Information Diffusion

Public education about the efficiencies of appliances is an important awareness measure to increase the pace of efficient devices. Information campaigns on the conservation of energy should be carried out through all major media. Potentially useful and effective programmes of economic incentives can be rendered ineffective because of lack of consumer awareness. These programmes must be integrated with policy tools that include information and education programmes and mandatory regulation of appliances. Guidelines to users and codes insisting that electric appliances bear labels that explicitly state the efficiency of their devices and the estimated operating costs. This would help consumers to make better informed decisions.

An information service cell for energy conservation could also be formed. Its duties would include: answering questions from the general public concerning energy conservation and energy-saving devices and making suggestions, offering technical assistance to scientists and engineers for developing energy-efficient devices. It should test the performance of various energy conservation materials and equipment. With all these measures, the cell can provide consumers with reliable data on the performance of different energy-conserving devices. In this context, endowment of technical manpower and Research and Development (R&D) institutions in an urban area have a major role to play in developing appropriate energy-efficient devices. In view of the long gestation periods associated with R&D, long-range planning to prepare for this technical development has to be included in the set of energy policies.

Energy and Planning: A Synthesis

This study has clearly established that urban areas are energy guzzlers contributing to the relative deprivation of energy elsewhere in the economy, especially in the case of high-quality and non-renewable energy carriers like petroleum products and electricity. Even wood-based fuels are renewable only when they are not used at a rate faster than that at which they are grown and produced. If so, the present pattern of usage of these fuels is most probably not sustainable. In addition to the environmental impact of deforestation, these carriers are also associated with large transportation energy as well as foreign exchange necessary for importing diesel for transportation. At present no policy instrument has been wielded to regulate the flow of these carriers. Restricting or escalating the energy supply tends to curb or stimulate metropolitan growth. City planners seem to have ignored this mechanism.

The per capita consumption of energy in an urban area is quite high when compared with rural areas as seen in this study. In future, one can expect *deepening of energy use* in a metropolitan city and *widening of energy use* in the rest of the economy. Factors which tend to cause the former are: increase in population, vertical and horizontal expansion resulting in intensive use of energy-intensive materials, changing the pattern of consumption of households, increasing the demand for transport, increasing importance of the service sector which has to exist where services are needed, increasing purchasing power due to increase in income levels, etc.

Therefore, the implication of energy consumption is a legitimate concern at all levels of urban planning. The extent to which energy considerations are integrated into various levels defines the policy implications. An agency can be formed involving local bodies, electricity boards, State and Central Government agencies, industries, consumer forums and members of the public to provide solutions to the energy problems of a city. This agency can study the present consumption and project energy demands, study alternative possibilities for supply (including renewables) and reduce the demand through conservation measures. It can also study the future land-use pattern, transport, etc. To achieve this goal, programmes of public education and technical and financial assistance for conservation are of vital importance. A broad range of policy options have to be studied and a select few can be finalised. The criteria for evaluation could be based only on efficiency. This framework could become an integral part of urban planning.

Some idea of the relative magnitude of energy requirement in policy alternatives together with an indication of their flexibility in relation to energy supply and demand can show whether they have significantly different energy implications and would be a great improvement on the present situation.

A restricted approach of considering the urban areas alone by guaranteeing the required total supply of energy cannot serve the ultimate goal of ensuring various energy services to the end-users of energy. One cannot assure end-use services by computations and planning through supply-based approaches alone. All the links in the fuel-cycle from primary to secondary to final to useful energy are to be safeguarded.

A two-stage approach is called for to avoid the possible slippages between the strategy and goals. The first step is to consider allocation of energy resources among the regions of the state or country, satisfying the equity criterion. Given the energy resources for a particular region, the second step would be to allot them optimally.

It is necessary to quantify energy resources which can be subject to freedom of allocation by the State or Central Government and those which do not fall in this category; it is also necessary to quantify resources in the first category which can be subject to regional allocation. Energy resources which can be allocated can be distributed among various regions in the State, based on criteria such as relative energy deprivation, development level, special needs, quantum of locally available energy resources outside the framework of allocation, with due consideration to prevention of over-exploitation. Energy resource availability in a particular region would be the sum total of the locally available resources and resources allocated from the *central pool*. Given energy resources in a particular region, the next step of the strategy involves specification of the optimum energy path.

Implementation of an energy plan raises a number of methodological and policy issues. Energy resource allocation across regions, on the basis of the criteria suggested, calls for the scrutiny and assessment of several components involved. It is necessary to take a view on the regions of the States to be considered. These need not correspond to political regions though considerations of an information system would compel one to opt for the same definitions. There is also the need to construct a Reference Energy System for the region, covering the phases of the fuel-cycle spelt out earlier. It is also necessary to understand the energy-economy interaction. These are essential steps for understanding the energy sub-system and its relation to the economy and are needed for going through both the stages of energy resource allocation.

Once the analysis gives an insight into the solutions, based on considerations of equity and efficiency, the next and more important step consists of spelling out policy parameters needed to achieve the desired energy resource allocation, structural changes in production and consumption and changes in production methods and technologies. This calls for a critical examination of efficacy of various policy instruments such as reliance on market forces, administrative allocation of energy carriers, capital and technology, tariffs and tax-based subsidies, incentives, penalties and disincentives, government-sponsored competitions, administrative generation of data and dissemination of information, administrative setting R&D priorities and support of R&D, and creation of an appropriate policy agenda. Without such a framework, planners are bound to fail in the implementation of energy-related policies.

Perhaps, the most fundamental question of energy policy is how the energy resources should be shared considering future generations. From this point of view, the development of renewable sources of energy is the ultimate solution. In the mix of such resources, solar energy for low-temperature heating and biomass in the form of energy forests distributed as a green belt around the urban centres are obvious components.

Future Perspectives

With a higher rate of economic growth, fuelled by economic reform, cities are likely to face further pressure for expansion, all of which will make major demands on its infrastructure including energy, telecommunications, transport, etc. Table 8.1 is a matrix depicting the influence of various parameters on future urban energy requirements.

The first factor to be considered is the population. With an increasing population, the quantity of energy carriers utilised also increases. Another important factor is income. With increasing incomes not only does the energy carrier consumption increase with respect to time, but the proportion of carriers allocated among different income categories also changes. Availability of carriers like petroleum products and electricity, and relative fuel prices cause the renewable energy sources to play an important role in future. With an almost explosive growth of urban centres, the distance between the home and workplace increases, thereby increasing transport energy consumption. The last factor is the efficiency of carrier utilisation. With an increasing dependence on non-renewables and environmental considerations, energy conservation and carrier substitution dictate the future energy consumption levels. As the above arguments show, in the near future, two factors will play an important role in the reduction of energy demand and one factor will contribute to its acceleration. The former are improvements in carrier utilisation and reduced dependence on non-renewables; the latter is an increase in population and income.

Consideration should be given to the following questions:

- How does the transport policy influence passengers in encouraging modal shifts and thereby vehicle kms?

Table 8.1: Future energy scenario.

Characteristics	Influencing parameters							
	Population	Income	Price of carrier	Investment in conservation	Government policies	Location and degree of dispersal		
						Households	Industries	Services
Energy requirement for household cooking and heating	*	*	*	*	*	*		
Energy requirement for industrial applications			*	*	*		*	
Energy requirement for transport	*	*	*	*	*	*	*	*
Potential for efficient residential technologies	*	*	*	*	*	*		
Potential for efficient industrial technologies			*	*	*		*	
Potential for efficient public transport system	*		*	*	*	*	*	*
Potential for renewable sources of energy	*		*	*	*	*	*	

Note: (*) = significant impact

- How will the change in the fuel prices affect the pattern of consumption of various carriers?
- How do efficiency improvements affect energy intensities in the industrial sector?
- To what extent do the energy carrier shifts affect future projections of various carriers?

Generally speaking, the main objective could be the estimation of energy for various end-uses sectors and allocation of resources accordingly. For future allocation, a range of parameters which influence the energy consumption pattern has to be taken into consideration.

The success of such urban policies in shaping the future of our cities and particularly the poor and under-privileged that depend on it to do a better job will be a function of our resolve to create new institutional arrangements of urban governance. For a long time, the urban areas have depended on the State as well as Central Government and non-government organisations (NGOs) (well-meaning but ineffectual). A different paradigm is essential to save the present degenerating urban energy systems.

In this approach, work could jointly be undertaken with representatives of State and Central Governments, NGOs and a panel of experts to upgrade such energy sectors as petroleum products, electricity, coal and renewable sources. Special committees could be formed for providing new and innovative ideas for a proper resource allocation and conservation, a research foundation to work on the feasibility of renewable energy utilisation and device efficiencies and a modicum of lobbying at the political level to ensure implementation. However, one thing that should be noted is that the lower end of the spectrum needs to be improved first.

This can act as a blue-print to other beleaguered metros where citizens realise the futility of leaving every thing to the government.

Urban Systems: Need for Sustainable Development

The policies of the Third World countries for the past few decades have been based on the assumption that development can be accelerated by just mimicking the Western countries, rather than finding approaches to local conditions and features. The results have been unpleasant: increasing urbanisation, diversion of funds to big cities, massive migration of rural population to urban areas, and increasing energy requirements resulting in deforestation, and consumption of precious non-renewable energy resources like coal and petroleum products. The bigger the city, the larger are its energy requirements particularly for high density fuels because of the economy of scale in transportation. *High-density living patterns can be sustained only by high-density fuels, which may be wood- or oil-based.* If high-density fuels become expensive and hard to obtain, high-density living in urban areas cannot be supported. The major sufferers will be the poorer sections of society. Hence, there can be no doubt about the kind of energy policies that would be rational for urban areas to adopt for their sustainable development. The essential ingredients of such policies comprise energy conservation and carrier substitution (of renewables for non-renewables (e.g. biomass for kerosene) or for other non-renewables (e.g. coal for oil)).

Urban areas, promoting rapid growth, are unaffected by the elementary truth that a growth which is not compatible with the environment and creates problems for its inhabitants in the long run is a doubtful blessing. Unless the proponents of sustainable development produce evidence of a marked injury to future generations, the growth rate will continue. As Schumacher (1983) has pointed out

> An environmental setting developed over millions of years must be considered to have some merit. Anything so complicated as a planet inhabited by more

> than a million and half species and animals, all of them living together in a more or less balanced equilibrium in which they continuously use and re-use the same molecules of the soil and air cannot be improved by aimless and uninformed tinkering. All changes in a complex mechanism involve some risk and could be made on a small scale first so as to provide a test before they are widely applied. When information is incomplete, changes could stay close to the natural process which have the undisputable evidence of having supported life for a very long time.

Urban systems do not seem to follow this golden rule. With their uninhibited appetite for growth, urban areas are completely changing the environmental setting thereby straining the future of mankind itself.

An urban policy based on sustainable development can surely restore a modicum of balance. The basic requirement of such a policy could be the avoidance of wasteful use of natural resources, which is synonymous with urbanisation, and a proper resource allocation. Only then can the environmental degradation, tying-up of transport capacity, unnecessary consumption of high-quality diesel fuel and its associated foreign exchange expenditure be avoided.

One should not forget that the solutions to energy problems of a metropolitan region are related to those for the rest of the region and *vice versa*. Hence, one cannot function in isolation though organisations are meant to make decisions regarding one region or the other. It is hoped that one can evolve an urban energy system which can sustain an economy and move on a development path for satisfying the basic human needs, at the same time ensuring *sustainability*.

REFERENCES

Alam, M., Dunkerley, J., Gopi, K.N., Ramsay, W., and Davis, E, 1983
Fuelwood survey of Hyderabad, Resources for the future, Washington. D.C.

Amara Pongsapich and Wathana Wongsekiarttirat, 1994
Urban household energy consumption in Thailand, Energy-*The International Journal*, Vol 19, No 5, 1994, pp 509-516.

Anantha Narayana, S (Forest Research Laboratory, Bangalore), 1988
Personal communication.

Anderson, K. P., 1973
Residential energy use: An econometric analysis, The Rand corporation (R-1297-NSF).

Anon, 1948
The Wealth of India: A Dictionary of Indian raw materials and industrial products, Council of Scientific and Industrial Research, New Delhi, India.

Anon, 1971
Encyclopedia of Energy, McGraw-Hill Book Company, New York.

Anon, 1974
Report of the Fuel Policy Committee, Planning Commission, Government of India.

Anon, 1975
Energy: Readings from Scientific American, W.H. Freeman and Company, San Fransisco.

Anon, 1978
Survey of rural energy consumption in Northern India, National Council of Applied Economic Research, New Delhi.

Anon, 1978
Energy handbook, Von Nostrand Reinhold Company, New York.

Anon, 1979
Report of the Working Group on Energy Policy, Planning Commission, Government of India, New Delhi.

Anon, (1980-90)
Annual Reports, Indian Oil Corporation, Bharat Petroleum Corporation and Hindustan Petroleum Corporation, Bangalore.

Anon, 1981
Indian Railways Year Book, Government of India, New Delhi.

Anon, 1982
Rural energy consumption pattern: A field study, ASTRA, Indian Institute of Science, Bangalore.

Anon, 1985
Macmillan Dictionary of Energy, Macmillan Press Ltd, London.

Anon, 1986
Energy in the cities of the Third World, Commission of the European Communities- Directorate General for Development, Brussels.

Anon, 1988
Annual Report, Beaureu of Economics and Statistics, Government of Karnataka, Bangalore.

Anon, 1988
Annual Report, Karnataka Power Corporation, Bangalore.

Anon, 1990
Annual Reports (1980-90), Karnataka Electricity Board, Bangalore.

Anon, 1990
Solar water-heating systems – Overview, Karnataka State Council for Science and Technology, Bangalore.

Anon, 1991
Census of India, Part I, Government of India Press, Bangalore.

Archibad R.B.,Finifter D.H., Moody, E., 1984
Seasonal variations in residential electricity demand: evidence from survey data, *Applied Economics*, Vol 14, 167-181.

Blackman, A.W.Jr, 1972
A mathematical model for trend forecasts, *Technological Forecasting and Social Change*, Vol 3, 441-452.

Baxter, R.E.,and Rees, R, 1978
Analysis of the industrial demand for electricity, *Economic Journal*, Vol 78, 277-298.

Barnes, R., Gillingham, R., and Hagemann, R, 1981
The short-run residential demand for electricity, *Review of Economics and Statistics*, Vol. 63, 541-552.

Beierlein, J.G., Dunn, J.W, and McConnon, J.C, Jr, 1981
The demand for Electricity and Natural Gas in the Northeastern United States, *Review of Economics and Statistics*, Vol. 63, 403-408.

Blakeley, P.W., 1974
Patterns of use of electrical energy in New Zealand, New Zealand Energy Conference.

Bodger, P.S., and Tay, H.S., 1981
Logistic and Energy Substitution Models for Electricity Forecasting: A Comparison Using New Zealand Consumption Data, *Technological Forecasting and Social Change*, Vol 31, 27-48.

Breese G (ed), 1969
The city in newly developing countries: readings in urbanism and urbanisation, Prentice Hall.

Bruce N, and Ellis G.M., 1993
Environmental Taxes and Policies for Developing Countries, Policy Research Department, The World Bank.

Castells, M., 1977
The urban question: a Marxist approach, Edward Arnold.

Clark, W., 1977
Energy for survival: The alternative to extinction, Allied Publishers Pvt.Ltd., New Delhi.

Chapman, D., Mount, T., and Timothy, T., 1972
Electricity demand growth and the energy crisis, *Science*, Vol 178, 703-708.

Clarke, C.G., 1974
Urbanisation in the Carrebbean, *Geography*, 59, 223-32.

Commerce Weekly, 1990
Vol 48, No 4050, Feb 19-25.

Dattatri, C and Siddarth Bhatt, M., 1985
Energetics of transportation systems, *Energy Management*, pp 135-152.

Daxiong Qiu, Yuquing Ma, Yingyun Lu, Qizhi Wang and Zhulinag Zhu, 1994
Household energy consumption in Beijing and Nanning, China, *Energy-The International Journal*, Vol 19, No 5, pp 529-538.

Dubin, J.A. and McFadden, D.L., 1984
An econometric analysis of residential electric appliance holdings and consumption, *Econometrica*, Vol 52, No 2, 345-362

Dunkerley, J., 1980
Domestic energy consumption by the poor in developing countries, in International energy studies, R.K. Pachuri (ed), Wiley and Sons, New Delhi.

Evers, A.D, 1975
Urbanisation and Urban conflict in Southeast Asia, *Asian Survey*, Vol. 15, 775-85.

Fisher, F.M and Kayson, C.A., 1962
A study in Econometrics: The demand for electricity in United States, North Holland Publishing Co., Amsterdam.

Fisher, J.C., and Pry, R.H., 1971
A simple substitution model of technological change, *Technological Forecasting and Social Change*, Vol 3, 75-88.

Garbacz, C., 1983
A model of Residential demand for electricity using a national household sample, *Energy Economics*,Vol 5, 124-128.

Garica, R.A., Manegdeg F and Raneses, N.O, 1994
Household energy consumption surveys in three Philipine cities, *Energy-The International Journal*, Vol 19, 539-548.

Gilbert A, and Gugler J, 1981
Cities, Poverty, and Development: Urbanisation in the third world, Oxford University Press, London.

Giuseppec and Ruggeri, 1983
Trends in household energy consumption in Canada, 1961-1980, *Energy policy*, Vol. 11, 250-258.

Goldemberg, J, 1981
Brazil: Energy options and current outlook, *Science*, Vol 200, 158-164.

Goldemberg, J, 1982
Energy issues and policies in Brazil, *Annual Rev. of Energy*, Vol 7.

Goldemberg, J, Johanson T.B., Reddy A.K.N., and Williams, R.H., 1988
Energy for a Sustainable World, Wiley Eastern ltd, Wiley Eastern Press, New Delhi.

Griffin, J. M., 1974
A long-term forecasting model of electricity demand and fuel requirements, *Bell Journal of Economics and Manamgement Science*, Vol 5, 515-539.

Halvorsen, R., 1975
Residential demand for electrical energy, *Review of Economics and Statistics*, Vol 57, 12-18.

Hardoy, J.E. (ed), 1975
Urbanisation in Latin America: approaches and issues, Anchor/Doubleday.

Hausman J.A., 1979
Individual discount rates and the purchase and utilisation of energy-using durables, *Bell Journal of Economics*, Vol 10, No 33, 33-54.

Henderson, P. D., 1975
India: The Energy Sector, Oxford University Press, New York.

Harker, J.H. and Backhurst J.R., 1981
Fuei and energy, Academic Press, London.

Herbert J.D and Van Huycak A.P., 1968
Urban planning in developing countries, Praeger.

Hermon, R., and Montroll, E.W., 1972
A manner of characterising the development of countries, *Proceedings of National Academy of Sciences*, Vol 69, 3019-3023.

Herric B.H, 1965
Urban migration and economic development in Chile, MIT Press.

Hills P., 1987
The Hongkong Energy Economy: A Review, *Pasific and Asian Journal of Energy*, Vol 1, No 1.

Hills, P., 1994
Household Energy Transition in Hongkong, *Energy-The International Journal*, Vol 19, 517-528.

Hirst, E., 1982
Residential energy use: Analysis of disaggregated data, *Energy Economics*, Vol 7, 74-82.

Hosier, R.H., 1986
Energy planning in Zimbabwe: An integrated approach, *Ambio*, Vol 15, 90-96.

Hosier, R., H., and Dowd, J., 1987
Household fuel choice in Zimbabwe: an empirical test of the energy ladder hypothesis, *Resources and Energy*, Vol 9, 347-361.

Hoiser, R.H., and Kepondya, W. 1993
Urban household energy use in Tanzania: prices, substitutes, *Energy Policy*, Vol 21, 454-473.

Houthakker, H. S., 1951
Some calculations on electricity consumption in Great Britain, *Journal of Royal Statistical Society*, Vol 114, 359-371.

Houthekar, H.S., Verleger, P.K., and Sheena D.P., 1973
Dynamic demand analyses for gasoline and residential electricity, Lexington, Mass:Data Sources, Inc.

Hughes Cromwick, E.L., 1985
Nairobi households and their energy use: An economic analysis of consumption patterns, *Energy Economics*, Vol 7, 265-278.

Ironmonger, D., Manning, I., and Von Hoa, T., 1984
Longitudinal Working Models: Estimates of household energy consumption in Australia, *Energy Economics*, Vol 6, 41-46.

Jain D.K, Satsangi P.S, and Kothari D.P., 1994
Urbanisation and energy-related problems in Indian cities, *Energy Sources*, Vol 16, 177-182.

Jayant Sataye and Steven Meyers, 1985
Energy use in cities of the developing countries (draft paper), Berkeley, CA.

Jones, D.W., 1991
How urbanisation affects energy use in developing countries, *Energy Policy*, Vol 9, 621-630.

Karnik, A.V., 1989
Energy in Indian Transportation: The emerging scenario, Himalaya Publishing House, New Delhi

Kulkarni, A., Sant G, Krishnayya, J.G., 1994

Urbanisation in search of energy in three Indian cities, *Energy-The International Journal*, Vol 19, No 5, pp 549-560.

Lipton M., 1977

Why poor stay poor: a study of urban bias in world development, Temple Smith/Harward University Press.

Maddala, S., 1986

Limited dependent and quantitative variables in Econometrics, McGraw-Hill Book Company, Singapore.

Mansfield, E., 1961

Technological change and the rate of immitation, *Econometrica*, Vol 29, 741-66.

Marcetti, C., 1972

Primary substitution models: on the interaction between energy and society, *Technological Forecasting and Social Change*, Vol 9, 77-84.

Marcetti, C., 1973

Primary substitution models, *Technological Forecasting and Social Change*, Vol 10, 345-356.

Marketing International Group, 1985

Energy Outlook, Elsevier Science Publishers, Amsterdam, The Netherlands.

McFadden, D., Puig, C. and Kirshner, D., 1977

Determinants of long-run demand for electricity, *Proceedings of the American Statistical Association*.

Meier A.K and Whitter J., 1983

Consumer discount rates implied by purchase of energy efficient refrigerators, *Energy*, Vol 8, 957-962.

Meyers, R.A.(ed), 1983

Handbook of Energy Technology and Economics, John Wiley and Sons, New York.

Misra P. P., (ed), 1978
Million cities of India, Vikas Publishing House, New Delhi.

Mohan R., 1979
Urban Economic Planning Models, The Johns Hopkins University Press, London.

Mount, T. D., Chapman, L.D. and Tyrell T.J., 1973
Electricity Demand in the United States: An econometric analysis, Oak Ridge National Laboratory, Oak Ridge, Tennessse.

Murphy, R., 1972
City and countryside as ideological issues: India and China, *Comparative Studies in History and Society*, Vol. 14, 250-67.

National Council for Applied Economic Research, 1958
Domestic fuels in India, Asia Publishing House, New Delhi.

Nicoles, G. and Pregogene, I., 1977
Self organisation in non-equilibrium systems: from dissipative structures to order through fluctuations, Wiley Inter Science Publications, John-Wiley and Sons, New York.

Owens, S., 1986
Energy, Planning and Urbanisation, Pion Ltd,London.

Pachuri, R.K., 1978
Energy consumption patterns: International energy studies, McGraw-Hill Publishing Company, New Delhi.

Panduri, B.P. *et al.*, 1985
Contribution of vehicles to air pollution in Indian cities, Paper presented at the Second World Congress on Engineering and Environment, November 7-9, New Delhi.

Parikh, J.K., 1980
Energy Systems and Development, Oxford University Press, New Delhi

Parikh, J.K., Sudhakara Reddy B., and Banerjee R., 1994
Planning for Demand Side Management in the Electricity Sector, Tata McGraw-Hill Publishing House, New Delhi.

Parikh K.S., 1976
Second India Studies: Energy, McMillan Co., New Delhi.

Pundir, B.P., and Sudkakar D., 1985
State of art: Report on vehicle emissions, Department of Environment, Government of India, New Delhi.

Parti, M., and Parti, C., 1980
The total and appliance-specific conditional demand for electricity in the household sector, *Bell Journal of Economics*, Vol 11, 309-321.

Paul, S., 1981
Biomass plantations - Energy planning, *Journal of Scientific and Industrial Research*, Vol 40, 67-78.

Pearl, Raymond, 1985
The biology of population growth, Alfred A.Knoff, New York.

Prakasa Rao,V.L.S., and Tewari, V.K., 1979
The Structure of an Indian Metropolis: A study of Bangalore, Allied Publishers, New Delhi.

Ramachandran, R, 1989
Urbanisation and Urban Systems in India, Oxford University Press, New Delhi.

Ramanathan R., and Sudhakara Reddy, 1994
Efficiencies of Energy Flow in Bangalore City: A micro level approach, *Energy Management (to appear).*

Reddy, A.K.N., and Krishna Prasad, K., 1977
Technological alternatives and the Indian energy crisis, *Economic and Political Weekly*, Vol 12, 1365-1402.

Reddy, A.K.N., et al, 1980
Rural energy consumption patterns: a field study, *Biomass*, Vol 2, 255-280.

Reddy, A.K.N., 1980
Alternative energy policies for developing countries: A case study of India, World energy production and productivity, (eds) Bohm, R.A.Clinard., and English, M.R., Ballingar Publications Co., Cambridge (Mass), USA, 289-351.

Reddy, A.K.N., 1982
Technology, development and the environment- a Re-appraisal, United National Environment Programme, Nairobi.

Reddy, A.K.N., 1986
Lecture notes on Energy Management for Development, Department of Management Studies, Indian Institute of Science, Bangalore.

Reddy, A.K.N., and Reddy, B.S., 1983
Energy in a stratified society: A case study of firewood in Bangalore, *Economic and Political Weekly*, Vol 17, 1757-70.

Reddy, A.K.N., and Reddy, B.S., 1986
The energy sector of the metropolis of Bangalore, Paper presented at the National Symposium on "The Planning and Development of Bangalore" at Bangalore.

Reddy, A.K.N., Sumitra, G., and Reddy, B.S., 1987
Energy requirements for 2000 A.D. Paper presented at the seminar on Bangalore 2000 - Some imperatives for action now, Bangalore.

Reddy, A.K.N., et.al., 1988
Karnataka Regional submodel and agricultural submodel, Project Report submitted to the Advisory Board on Energy, Government of India.

Reddy, A.K.N., Sumithra, G.D., Balachandra, P. and D'sa A., 1990
Comparative costs of centralised and decentralised electricity generation, *Economic and Political Weekly*, Vol. XXV, 1201-1216.

Reddy, A.K.N., Sumithra, G.D, Balachandra, P. and D'sa A., 1991
A Development Focused End Use Oriented Electricity Scenario for Karnataka, *Economic and Political Weekly*, Vol.XXVI, 891-910.

Reddy, A.K.N., Sumithra, G.D., Balachandra, P., and D'sa A., 1991
A Development Focused End Use Oriented Electricity Scenario for Karnataka, *Economic and Political Weekly*, Vol. XXVI, 983-1001.

Reddy, A.K.N., 1991
Barriers to improvements in energy efficiency, *Energy Policy*, Vol 19, 953-961.

Reddy, A.K.N., and Reddy B.S., 1994
Substitution of energy carriers for cooking in the metropolis of Bangalore, *Energy - the International Journal*, Vol 19, 561-571.

Rogers, A., and Willamson G.W., (editors), 1982
Urbanisation and Development in the Third World, International Institute for Applied Systems Analysis, Lexenburg, Austria.

Romn, J., 1973
Urbanisation in Thailand, Ford Foundation.

Ruedisili, L.C., and Morris, W.F., 1975
Prospectives on energy: Issues, ideas and environment, Oxford University press, New York, 1975.

Ruderman H., Levine M., and McMohan J., 1987
Energy-Efficiency choice in the purchase of residential appliances, in *Energy Efficiency: Perspectives on individual behaviour*, American Council for Energy Efficient Economy, Washington, D.C., Berkeley, CA.

Ruggeri, G.C., 1973
Trends in household energy consumption in Canada, 1961-1980, *Journal of Business*, 526-553.

Sanders, D.H., 1980
Statistics: A fresh approach, McGraw-Hill, Kogukusha, Tokyo.

Schipper, L., 1983
Residential energy use and conservation in Denmark, 1965-1980, *Energy Policy*, Vol. 11, 313-322.

Sathaye J, Tyler S., and Goldman N., 1994
Transportation, fuel use and air quality in Asian cities, *Energy-The International Journal*, Vol 19, 573-586.

Schumaker, 1973
Small is beautiful, Blond and Briggs, London.

Geoffrey, K., (ed) 1983
Schumaker on Energy, Abacus, London.

Singer, P., 1975
Urbanisation and development:the case of Sao Paulo, in Hardoy (ed), 435-56.

Sharif, M.N., and Islam, M.N., 1980
The Weibull Distribution as a general model for forecasting technological change, *Technological Forecasting and Social Change*, Vol 18, 247-256.

Smith K.R., Apte M.G., Yuqing Ma, Wongsekiattirat and Kulkatni A., 1994
Air Pollution and the energy ladder in Asian cities, *Energy-The International Journal*, Vol 19, 587-600.

Sreehari, M.N., 1987
Transportation planning for Bangalore Metropolis: A scenario for 2001, Paper presented at the "National Symposium on Planning and Development of Bangalore", Bangalore.

Steinhart, C.E., and Steinhart, J.S., 1974
Energy: Sources, use and role in human affairs, Duxbury Press, Northscitute, Massachusetts.

Sudhakara Reddy, B., 1990
The Energy Sector of the Metropolis of Bangalore, Ph.D thesis, Indian Institute of Science, Bangalore.

Sudhakara Reddy, B., 1995
The concept and evaluation of Demand Side Management, *International Journal of Global Energy Issues,* Vol 7, 48-55.

Sudhakara Reddy, B, 1995

Appliance electricity consumption in the residential sector - An economic approach, *Energy Sources*, Vol 17, 179-193.

Sudhakara Reddy, B., 1995

Transportation, Energy and Environment: A case study of Bangalore, *Economic and Political Weekly*, Vol XXX.

Sudhakara Reddy, 1995

Consumer discount rates and energy carrier choices in urban households, *The International Journal of Energy research* Vol. 20, 187-195..

Sudhakara Reddy, B., 1995

An econometric analysis of energy use in Urban Households, *Energy Sources*, Vol 17, 359-372.

Sudhakara Reddy, B., 1995

"A multilogit model for fuel shifts in the residential sector", *Energy-The International Journal*, Vol. 20, 929-936.

Sudhakara Reddy, B., 1995

"Modelling of competing technologies in the residential sector", *Energy Conversion and Management*, Vol. 37, 117-125.

Sumitra, G, 1981, An analysis of 1977-78

National Sample Survey on household energy consumption, (32nd round) – Karnataka, Perspective Planning Division, Planning Dept, Govt of Karnataka.

Taylor, L, D., 1975

The demand for electricity: a survey, *Bell Journal of Economics*, Vol 6, 74-110.

Thornadike, E.H., 1978

Energy and Environment: A primer for scientists and engineers, Addison Wesley Publishing Company, New York.

Train K., 1985

Discount rates in consumers' energy related decisions: A review of literature", *Energy*, Vol 10, 1243-1253.

Turner J.F.C., 1969
Uncontrolled urban settelments: Problems and policies, in Breese (ed), 1969, 507-31.

Turner, R., (ed), 1962
India's urban future, University of California Press.

Tyter, G.H., 1986
Standard Handbook of Engineering Calculations, McGrawh-Hill Internationaledition, Singapore.

United Nations, 1989
Urban Transport Development with Particular Reference to Developing Countries.

Venkata Raman, L., 1988
Personal communication, Hindustan Petroleum Corporation Ltd Bangalore.

Veziroglu, T.,N., 1978
Alternative energy sources: An international compendium, Hemisphere Publishing Corporation, Washington.

Whitelegg, J., 1991
Transport, environment and energy, *Impact of Science on Society*, pp 89-100.

Wilder, R.P., and Willenborg, J.F., 1975
Residential demand for electricity: a consumer panel approach, *Southern Economic Journal*, Vol 42, 112-217.

Wilson, C., 1980
Energy for survival: The consumption of energy in an American society, Allied Publishers Pvt. Ltd, Bombay.

Wilson, J.W., 1974
Electricity consumption: supply requirements, demand elasticity and rate design, *American Journal of Agriculture of Economics.*

Wilson, J.W., 1971
Residential demand for electricity, *Review of Economic Business*, Vol 11, 7-19.

Wirth, L., 1938
Urbanisation as a way of life, *American Journal of Sociology*, 44, 1-24.

Zaki, A.S., and Isakson, R., 1983
The impact of energy prices, *Energy Economics*, Vol 5, 100-108.

INDEX